William Henry Flower

The Horse

A Study in Natural History

William Henry Flower

The Horse
A Study in Natural History

ISBN/EAN: 9783744662086

Printed in Europe, USA, Canada, Australia, Japan

Cover: Foto ©Andreas Hilbeck / pixelio.de

More available books at **www.hansebooks.com**

THE HORSE

A STUDY IN NATURAL HISTORY

BY

WILLIAM HENRY FLOWER, C. B.

LL. D., D. C. L., Sc. D., F. R. S., Pres. Z. S., Etc.

DIRECTOR OF THE BRITISH NATURAL HISTORY MUSEUM
FORMERLY HUNTERIAN PROFESSOR OF COMPARATIVE ANATOMY AND PHYSIOLOGY
AT THE ROYAL COLLEGE OF SURGEONS OF ENGLAND
AND SOMETIME EXAMINER IN ANATOMY
AT THE ROYAL COLLEGE OF VETERINARY SURGEONS

NEW YORK
D. APPLETON AND COMPANY
1892

DEDICATED

TO THE MEMORY OF MY FATHER

EDWARD FORDHAM FLOWER

(b. 1805, d. 1883)

WHO DEVOTED A GREAT PART OF HIS LIFE TO THE ENDEAVOUR

TO ALLEVIATE THE SUFFERINGS INFLICTED BY MAN

UPON THE HORSE

EDITOR'S INTRODUCTION.

THE works to be comprised in this Series are intended to give on each subject the information which an intelligent layman might wish to possess. They are not primarily intended for the young, nor for the specialist, though even to him they will doubtless be often useful in supplying references, or suggesting lines of research.

Each book will be complete in itself, care, however, being taken that while the books do not overlap, they supplement each other; and while scientific in treatment, they will be, as far as possible, presented in simple language, divested of needless technicalities.

The rapid progress of science has made it more and more difficult, and renders it now quite impossible, to master the works which appear, almost daily, on various branches of science, or to keep up with the proceedings of our numerous Scientific Societies.

A distinguished statesman has recently expressed the opinion, that we cannot expect in the next fifty years any advance in science at all comparable to that of the last half-century. Without wishing to dogmatise, I should be disposed to hope that in the future the progress of science will be even more rapid.

In the first place, the number of students is far greater; in the second, our means of research—the microscope and telescope, the spectroscope, photography, and many other ingenious appliances—are being added to and rendered more effective year by year; and, above all, the circle of science is ever widening, so that the farther we advance the more numerous are the problems opening out before us.

No doubt there are other Scientific Series, but it is not believed that the present will exactly compete with any of them. The *International Scientific Series* and *Nature Series* are no doubt useful and excellent, and some of the volumes contained in them would well carry out the ideas of the Publishers, but, as a rule, they are somewhat more technical and go into minuter details.

The names of the Authors are a sufficient guarantee that the subjects will be treated in an interesting and thoroughly scientific manner.

High Elms, Farnborough:
  November, 1891.

PREFACE

ACCORDING to Huth's valuable "Bibliographical Record of Hippology" (Works on Horses and Equitation), published in the year 1887, there had been up to that date at least 3,800 separate works published in the various languages of the civilised world on subjects appertaining to the horse. This enumeration is, of course, far from perfect, and very many additions are being made yearly to the list.

I confess that I should have felt some hesitation at adding another to this formidable array if I did not believe that the subject has never been approached from the standpoint of this little work, and that therefore something will be found in it which is not to be met with in any of those hitherto written. In fact, it is only the knowledge which has in very recent times accumulated from various sources which could have made such a work possible.

It endeavours to look at the horse as the animal appears in the light of the modern and now generally accepted doctrines of Natural History, and in thus doing it may be the means of teaching what some of those doctrines are, and so of affording insight into the methods of nature applicable to a far wider range of study and of thought than that limited to any single species.

By permission of the publishers of the "Encyclopædia Britannica" some passages from my articles on the Horse and allied animals which appeared in the ninth edition of that work have been incorporated in this memoir, and I am greatly indebted to Mr. Gambier Bolton, Major J. Fortuné Nott, and Mr. York, for the use of the original photographs from which the figures of the tapir, rhinoceros, and various members of the horse family have been reproduced. That of the quagga is especially interesting, as being from the only photograph known to have been taken of this animal in a living state.

W. H. F.

May, 1891.

CONTENTS

CHAPTER I.

Interest of the study of the horse, especially as illustrating some important principles in biology—A test case of the value of the theory of transmutation of species—Significance of rudimentary structures—Meaning of the term "specialization"—Position of the horse in the animal kingdom—Division of ungulate mammals into perissodactyle and artiodactyle—The horse belongs to the former—Palæontological history of the perissodactyles—Generalized ungulates of the earliest Eocene age—Phenacodus—True perissodactyles—Hyracotherium—Palæotherium—Families which became extinct without leaving descendants—Three surviving families, represented at the present time by the Tapirs, Rhinoceroses, and Horses—The first the least and the last the most modified — Principal characters by which horses differ from the generalized early forms of perissodactyles, probably all adaptations to changed conditions of life —Present state and probable future of the group.

CHAPTER II.

The tapirs (Family *Tapiridæ*)—Characters, species, geographical and geological distribution—The rhinoceroses (Family *Rhinocerotidæ*)—The horses (Family *Equidæ*)— Their immediate predecessors—The hipparions, or three-

ILLUSTRATIONS

1

THE HORSE.

CHAPTER I.

THE HORSE'S PLACE IN NATURE—ITS ANCESTORS AND RELATIONS.

Interest of the study of the horse, especially as illustrating some important principles in biology—A test case of the value of the theory of transmutation of species—Significance of rudimentary structures—Meaning of the term "specialization"—Position of the horse in the animal kingdom—Division of ungulate mammals into perissodactyle and artiodactyle—The horse belongs to the former—Palæontological history of the perissodactyles—Generalized ungulates of the earliest Eocene age—Phenacodus—True perissodactyles—Hyracotherium—Palæotherium—Families which became extinct without leaving descendants—Three surviving families, represented at the present time by the Tapirs, Rhinoceroses, and Horses—The first the least and the last the most modified—Principal characters by which horses differ from the generalized early forms of perissodactyles, probably all adaptations to changed conditions of life—Present state and probable future of the group.

THE horse is from many points of view one of the most interesting of animals. In utility to man it yields to no other. It was his domestic companion, friend, and servant before the dawn of history. It has accompanied him in his wanderings over almost every part of the surface of the earth, performing duties both in peace and war which no other animal could have done, and giving Man facilities for the exercise of dominion over nature which otherwise would have been impossible to him. The *rôle* of the ass, the ox, the camel, and the llama in performing similar duties has been of a limited and subsidiary nature compared to that of the horse.

It is only in very recent times that the progress of mechanical invention has begun to supersede some of the uses for which the strength and the speed of the horse for many thousands of years have alone been available. How far this commencing disestablishment of the horse from its unique position as the main agent by which man and his possessions have been carried and drawn all over the face of the earth will go, it is difficult to say at present.

To the eye of the naturalist, the horse presents other and still higher sources of interest. No better example can be found in the whole range of the animal kingdom to illustrate certain great principles found acting universally in the construction of the bodies of all living beings, whether animals or plants. The structure of the horse in relation to that of allied animals and to the actions which it has to perform in the economy of nature may be most advantageously studied by every one who wishes to gain an insight into some of the fundamental principles of biology. In scarcely any other animal has *specialization* of various parts—that is, modification from the general or average type to conform to the requirements of some special mode of existence—been carried to such an extreme. In many organs, but especially in the limbs and teeth, we find the strongest evidence of two opposing principles striving against each other for the mastery in fashioning their form and structure. We find *heredity*, or adherence to a general type derived from ancestors, opposed by special modifications of or deviations from that type, and the latter generally getting the victory, although in the numerous rudimentary structures that remain there is significant evidence of ancestral conditions long passed away. The various specializations, evidently in adaptation to purpose, will be thought by many to be the result of the survival, in the severe struggle for existence, of what is best fitted for the purpose to which it is to be applied. This may or may not be the explanation, but the interest of the study of such an animal as the horse will be increased tenfold by the conviction that there is some true and probably discoverable causation for all its modifications of structure, however far we may yet be from the true solution of the methods by which they have been brought about.

The anatomy and history of the horse are, moreover, often taken as affording a test case of the value of the theory of evolution, or, at all events, of the doctrine that animal forms have been transmuted or modified one from another with the advance of time, whether, as extreme evolutionists hold, by a spontaneous or inherent evolving or unrolling process, or, as many others are disposed to think, by some mysterious and supernatural guidance along certain definite lines of change. It will be observed that both these views are opposed to the doctrine, formerly held universally by naturalists and theologians alike, that each modification of animal or plant form sufficiently distinct to be called a species had a separate origin—a doctrine for which, it may be remarked by the way, no proof of any kind has ever been offered.

The evidence in favor of the theory of transmutation afforded by the case of the horse is derived from two distinct sources—(1) The structure of existing horses; (2) the past history of the race as revealed by fossil remains.

(1) By far the most interesting portions of the organization of existing horses from this point of view are the various rudimentary and apparently useless structures which occur in several parts of its body,

structures which correspond to some which are fully developed and functional in other animals, but which, in the horse, are so reduced in size or altered in character as to be of little or not any use in its economy.

Parts, usually called rudimentary, may be in one of two conditions: either *nascent,* or in process of growth to something larger and more useful; or *vestigial*—that is, in a dwindling and degenerate state, vestiges of a once more developed condition. In any particular case, it may be difficult to say to which category it should be assigned, and we may have to look for guidance beyond the mere structure itself. In all or nearly all which we shall meet with in the horse, the presence of the same parts in a fully developed state in other allied though less specialized animals points clearly to the second condition, a conclusion which is strengthened by the certain knowledge derived from palæontology that the horse in its present form has only come into existence at a very late period of the world's history—is, in fact, one of the most modern forms of animal known.

In tracing the history and affinities of animals, rudimentary organs are looked upon by naturalists as far more important than highly developed or functional parts. As Darwin says, they "may be compared with the letters of a word, still retained in the spelling but become useless in the pronunciation, but which serve as a clue for its derivation.* On the view of descent with modification we may conclude that the existence of organs in a rudimentary, imperfect, or useless condition, or quite aborted, far from presenting a strange difficulty, as they assuredly do on the old doctrine of creation, might even have been anticipated in accordance with the views here explained."

The rudimentary parts met with in the structure of the horse will be described fully in the last two chapters of this work, which treat of the anatomical characters of the animal.

(2) It is, however, to the ancestral history, as disclosed by palæontology, or the study of fossil remains, that we must look for the more direct evidence of the truth of the theory; and we are in a better position to do this in the case of the horse than in that perhaps of any other animal, as it is one of the few whose history can be traced through a tolerably complete chain of links as far back as the earliest Tertiary age.† We must, however, not carry

* As, for example, the *b* in "debt" and "doubt."
† The latest of the three great periods into which geologists divide the age of the earth is called Tertiary or Cainozoic. It is subdivided into Eocene, Miocene, Pliocene, and Pleistocene, the last being that which immediately preceded the one in which we are now living.

away the idea that the record is yet perfect. Before the commencement of the Eocene period it is wrapped in what appears at present impenetrable darkness and mystery.

Throughout the vast Tertiary period, fragments here and fragments there stand out among the ruins, from which we endeavor to reconstruct our edifice, just as the skillful architect or antiquary, from the shattered pieces of marble or stone of an ancient temple, will restore to us the noble forms and proportions it once bore.

The outcome of all recent work in this subject has been, that every fresh discovery which has been made has tended to corroborate, and nothing has been found inconsistent with, the view that the living beings which we see around us have been gradually fashioned into shape by the modification of pre-existing forms—a view of creation which is the grandest, most sublime, and at the same time most reasonable, which has yet been presented to us.

A few words may be said here upon the important subject of *specialization*, which will be so frequently referred to in what follows. The modifications in animal structure which come under this definition may be grouped under three principal headings: (1) The addition of parts not met with in the generality of animals, and, as far as is known, not

2

found in the earliest members of the group which afterwards possess them—as, for example, the antlers of deer, the horns of oxen or the rhinoceros, the humps of camels, etc.   (2) The suppression of parts commonly present—as the upper front teeth of ruminants, the tails of bears and guinea-pigs, the outer toes of the horse's foot, the entire hind limbs of porpoises, etc.   (3) The modification of the form, size, and relation of parts—as the immense development of the tusks in the walrus and male musk-deer, the complicated foldings of the grinding teeth of elephants, etc.

In tracing out any series of gradual modifications following each other in a regular chronological sequence, as we are sometimes fortunate enough to be able to do,* we find that progress is usually from the general to the special.   It must not, however, be supposed from this statement that all animals living in ancient times were more generalized in character than many now existing.   On the contrary, many of the extinct forms, even those of quite early periods, were in some portion of their structure very highly specialized.   In fact, high specialization almost invariably leads ultimately to extinction, be-

* Many such instances are described in an interesting series of works, entitled *Les Enchaînements du Monde Animal dans les Temps Géologiques*, by Professor Albert Gaudry, Paris, 1878–90.

cause it results from adaptation to particular conditions, which may become changed in course of time, and then the animals which have become adapted exclusively for life under those conditions perish, while those animals that retain more general characters readily adapt themselves to the altered circumstances. The commonplace, average sort of creatures are thus often the longest lived as species, while such very strangely modified forms as *Uintatherium*,* *Machairodus*,† and *Thylacoleo*,‡ passed rapidly over the stage and then vanished from sight.

It is proposed in this little work to treat of the horse, not as an isolated form, but as one link in a great chain, one term in a vast series, one twig of a mighty tree; and to endeavor to trace, as far as our present knowledge permits, what its relations are to the rest, and by what steps of modification in its

* A huge beast from the Eocene of North America, with limbs resembling those of an elephant, and a rhinoceros-like skull, but with great descending flattened tusks in the upper jaw, and three pairs of bony prominences, like horns, on the top of the head.

† An animal allied to the tiger, with enormous saber-like upper canines, found in the later Tertiaries of both Europe and America.

‡ A marsupial of the late Tertiary period of Australia, as large as a sheep, allied to the phalangers and kangaroos, but with one huge cutting cheek-tooth (premolar), and one great incisor on each side of each jaw, all the other teeth being extremely reduced in size and almost functionless.

various parts it has come to be the very singular and highly specialized animal we have now before us, so distinct from all existing forms of life that in most of the older zoological systems it was (at least asso-ciated only with some very immediate allies, struc-turally almost identical) placed in an order apart from all other mammals, under the name of *Solid-ungula*, *Solipedia*, or *Monodactyla*, the animal with the solid foot, or rather with a single toe on each ex-tremity.

As will be seen from the sequel, the various forms of asses and zebras only differ from the horse in slight details of their organization, and with it form a group entirely apart from all other existing animals, a group constituting the genus *Equus* and the family *Equidæ*, but no longer considered so isolated as to form a distinct order. In much of what follows the term "Horse," unless the contrary is especially stated, must be understood to include the other members of the family.

To understand the natural place of the horse in the zoological system it will be necessary to take a wide glance at the whole great group to which it belongs. That it is a vertebrate animal, and that it occupies a place in the class Mammalia, no one will doubt. Within that class there can also be no doubt about its taking its place in the great division of *Eutheria*, which includes all existing mammals except the Marsupials and Monotremes. In treating only of existing mammals, a division of the class into distinctly circumscribed groups is perfectly easy. The so-called *orders* generally accepted are separated from each other by well-marked breaks of continuity. Many extinct forms can also be contained within the definitions of these orders. But the rapid advance of palæontology is disclosing to view an enormous number of long-buried animal forms, which are gradually filling up all the spaces left between the isolated groups now surviving on the earth, and continually increasing the difficulty of giving satisfactory definitions of their boundaries.

In the first serious attempt at the classification of the Mammalia, that of John Ray, in his "Synopsis Methodica Animalium," published in 1693, the class was separated into two great divisions, the ungulated or hoofed animals, and the unguiculated, or those with nails or claws. This division, especially as applied by its author, was somewhat artificial, the camel being separated from all its ungulate allies, and placed in the unguiculate division, and the latter embracing a very heterogeneous assemblage of creatures. Nevertheless, some portion of this system has survived, and especially the group *Ungulata*, discarded by Linnæus, Cuvier, and others, and broken up by them in several distinct orders, has been resuscitated of late years, and is now generally used, with almost the same limits as were assigned to it by Ray.

The *Ungulata* in this sense are all animals eminently adapted for a terrestrial life, and in the main for a vegetable diet. Though a few are more or less omnivorous, and may under some circumstances kill living creatures smaller and weaker than themselves for food,* none are distinctly and habitually predaceous. Their molar or cheek-teeth have broad crowns, with tuberculated or ridged grinding surfaces, and they have a very completely developed set of milk-teeth, which are not changed until the animals have nearly attained maturity. Their limbs are adapted for carrying the body in ordinary terrestrial progression, and are of very little use for any other purpose, such as flying, climbing, seizing prey, or carrying food to the mouth. They have no clavicles or collar-bones. Their toes are provided with blunt, broad nails, which in the majority of cases more or less surround and inclose their ends, and are called hoofs. Leaving aside certain forms which are not so nearly related to the subject of this memoir as to concern us further and which are nearly all extinct, the majority of the ungulated animals have been

* Pigs, for instance, will kill and eat snakes.

throughout the whole of the Tertiary period separated into two perfectly distinct sections, differing from each other not only in the obvious characters of the structure of their limbs, but in numerous important points in other portions of their organization, such as their skull, vertebral column, teeth, digestive organs, etc. The characters of these two groups, first indicated by Cuvier, were thoroughly established by Owen, by whom the names by which they are now generally known were proposed. These are *Artiodactyla*, or even-toed, and *Perissodactyla*, or odd-toed.*

It is only by studying the fundamental type of organization common to all members of a group which underlies the various external or superficial modifications by which it becomes adapted to the different surrounding conditions under which it has to carry on its existence, that the true relationship of animals can be determined. In this way it can be clearly demonstrated that the pig, the deer, the ox, sheep, goat, antelope, and camel, including even such extreme forms as the giraffe and the hippopotamus, are formed on one plan—the Artiodactyle; while the horse, the tapir, and the rhinoceros are formed on the other—the Perissodactyle type.

* From the Greek *artios*, even in number, and *perissos*, uneven; combined with *daktylos*, finger or toe.

To understand one of the principal distinctions between these two forms, it must be premised by way of explanation that the number of digits (a convenient common term to express either fingers or toes, whether of the fore or hind foot) in mammals never exceeds five on each limb. For convenience of description, they are designated numerically from the inner side of the limb—I., II., III., IV. and V. (Fig. 1, p. 15) the pollex (thumb) and hallux (great toe) being the first of the fore and hind limbs respectively, and the third is the middle of the complete series. When the number falls short of five, it is always easy to determine, by their relations to the bones of the wrist or ankle, which of the typical series are present and which are missing.

In the Artiodactyles the third and fourth digits of both feet are almost equally developed, and flattened on their inner or contiguous surfaces, so that each is not symmetrical in itself, but when the two are placed together they form a figure symmetrically disposed to a line drawn between them, and constitute the erroneously called "cloven hoof" of the Ruminants, which is really not one, but the two hoofs of distinct toes. These two digits are always present and well developed; the second and fifth may be present in varying degrees of development, or may be entirely absent; the first is not present in any known member of the group, even the most ancient.

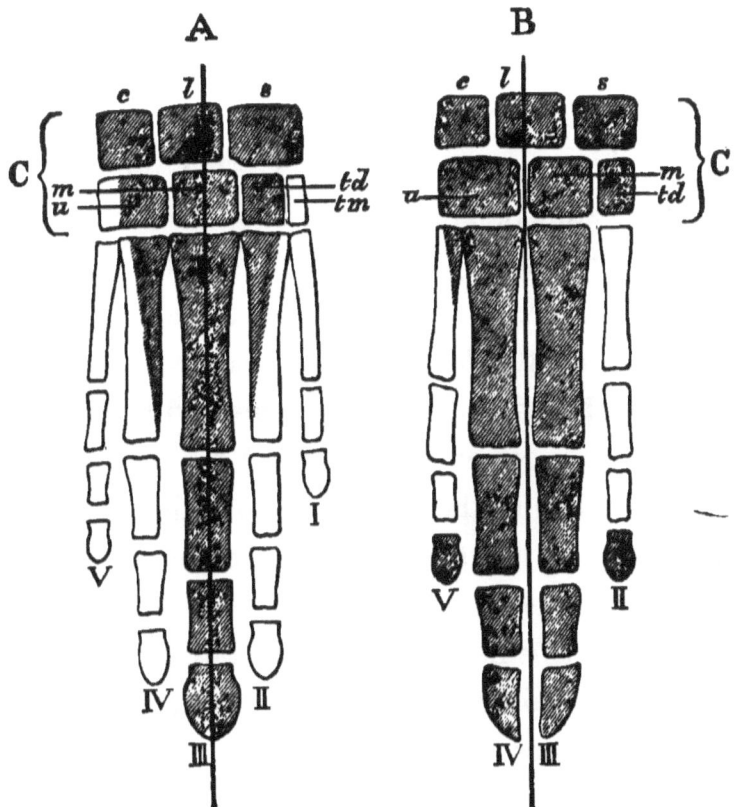

Fig. 1.—A, diagrammatic representation of the bones of the right fore-foot of an odd-toed or Perissodactyle animal. B, of an even-toed or Artiodactyle. C, the carpus or wrist, consisting of two rows of bones, the upper being *c*, cuneiform; *l*, lunar; and *s*, scaphoid; the lower, *u*, unciform; *m*, magnum; *td*, trapezoid, and *tm*, trapezium. The long bones in contact with the last constitute the metacarpus; the remaining bones are the phalanges. The digits or toes are numbered in order from the inner to the outer side of the foot. The shaded parts of A are those that are present in the horse; in B, those that are present in the ox.

In the Perissodactyle group, the middle or third digit of both fore and hind feet is larger than any of the others, and symmetrical in itself, the free border of the last bone (the ungual phalanx, which supports the hoof *) being evenly rounded on both sides. This may be the only digit sufficiently large to be of any use to the animal, as in the horse, or the second and fourth may be nearly equally developed on each side of it, as in the rhinoceros. In the tapir and in many extinct forms, the fifth digit is also present in the fore foot; but this does not interfere with the symmetrical arrangement of the rest of the foot around the median line of the third or middle digit. A first digit (pollex or hallux) has only been found in some extremely ancient and primitive forms.

It will not be necessary to enter into any description of the remaining anatomical characters by which these two groups are distinguished, although, as before said, they are very marked, and pervade almost every portion of their organization. The differen-

---

* The phalanges are the separate bones of which the digits are composed. They are three in number in each digit, called respectively first, second, and third, or proximal, middle, and distal; the last being often also called "ungual," because it supports the nail or hoof. The metacarpals and metatarsals are the long bones which connect the carpus (wrist-bones) or tarsus (ankle-bones) with the digits in the fore and hind limb respectively. When a word common for both is required, they are spoken of as "metapodials."

tial characters of the feet can be readily appreciated, even by those who have little anatomical knowledge, and suffice to show the fundamental distinction between them.

Having now eliminated from consideration all mammals but ungulates, and all ungulates but perissodactyles, we may henceforth confine our attention solely to this group, as it is the one which contains the horse and all its nearest relatives, and we must first endeavor to trace its history back in geological time as far as our available records will take us.

It is now well known that mammals existed far back into the secondary or mesozoic age, as far back as the Rhætic or uppermost beds of the Triassic system; but these had none of the characters of ungulates. They were all very small in size, and apparently more nearly allied to the Marsupialia and Insectivora than to any other existing orders. Until quite recently not a trace of any mammal had been found in any of the strata attributed to the great Cretaceous epoch. The blank has, however, been partially filled up by the discoveries in North America announced by Professor Marsh; but we know as yet too little of these to be able to form any satisfactory opinion as to their affinities, or to pronounce with any certainty whether they carry back the pedigree of the perissodactyle group beyond the commencement of the Tertiary period. At present the balance of evidence is rather in favor of their relationship with the earlier and more primitive forms just mentioned. We have, however, certain knowledge that when the land which formed the bottom of the great cretaceous ocean which flowed over a considerable part of the present continents of Europe and North America was lifted above the level of the water and became fitted to be the abode of terrestrial animals, it was very soon the habitation of vast numbers of herbivorous and hoofed mammals.

The remains of animals to which it is possible to trace back the modern horse by a series of successive modifications without any great break are found in abundance in the lower strata of the great lacustrine formations assigned to the Eocene period spread over considerable portions of the present territories of New Mexico, Wyoming, and Utah, in North America. Similar animals also existed in other parts of the world, but in Europe the hitherto-discovered fragments which prove their existence are in a less complete and satisfactory condition for investigation. Negative evidence is in such cases, however, of little value, as may be judged from the fact that it is only within a very few years that the existence of these American deposits teeming with fossil remains of previously unsuspected forms of life has been brought to light.   How do we not know that the next ten or twenty years may not be equally fruitful in new discovery?

After giving a summary of what was then known of the ancestry of the horse, as disclosed by palæontological evidence, Professor Huxley wrote in 1877 :* "The knowledge we now possess justifies us completely in the anticipation that when the still lower Eocene deposits and those which belong to the cretaceous epoch have yielded up their remains of ancestral equine animals, we shall find, first, a form with four complete toes and a rudiment of the innermost or first digit in front, with probably a rudiment of the fifth digit in the hind foot; while in still older forms, the series of the digits will be more and more complete, until we come to the five-toed animals, in which, if the doctrine of evolution is well founded, the whole series must have taken origin."

This anticipation has been completely verified by the discovery, among others, of *Phenacodus* in the Wasatch beds, which there is every reason to believe are nearly, if not quite, the oldest of the Eocene formations of North America.

Although this most interesting animal was known and named by Cope as long ago as 1873 from teeth alone, it was not until the more recent discov-

American Addresses : Lectures on Evolution, p. 89.

eries by Wortman of complete skeletons of more than one individual with all their bones in connection that we were put in possession of almost as perfect a knowledge of its osteological characters as of any animal now existing. The figures and descriptions published by Professor Cope,* and the excellent casts sent to this country of one of the skeletons, have made this knowledge widely accessible. Although this creature was of an extremely generalized form, it was obviously so far separated from the primitive mammalian type, whatever that may have been, as to come within the definition of the ungulate group, using this term in its widest sense. The terminal bones of the toes were of such a form as to show that they were incased in hoofs, instead of carrying claws, and it had no clavicles. The teeth also were adapted for a herbivorous or omnivorous diet.

Phenacodus, however, does not stand alone, even in our present state of knowledge; it belongs to a family of which several generic modifications are already described, and remains of still more generalized forms, the *Periptychidæ* of Cope, are found in the Puerco Eocene beds of New Mexico, probably still older than the Wasatch. Forms apparently allied have also been discovered by Rütimeyer in

---

* *Report of the United States Survey of the Territories*, vol. iii. 1884.

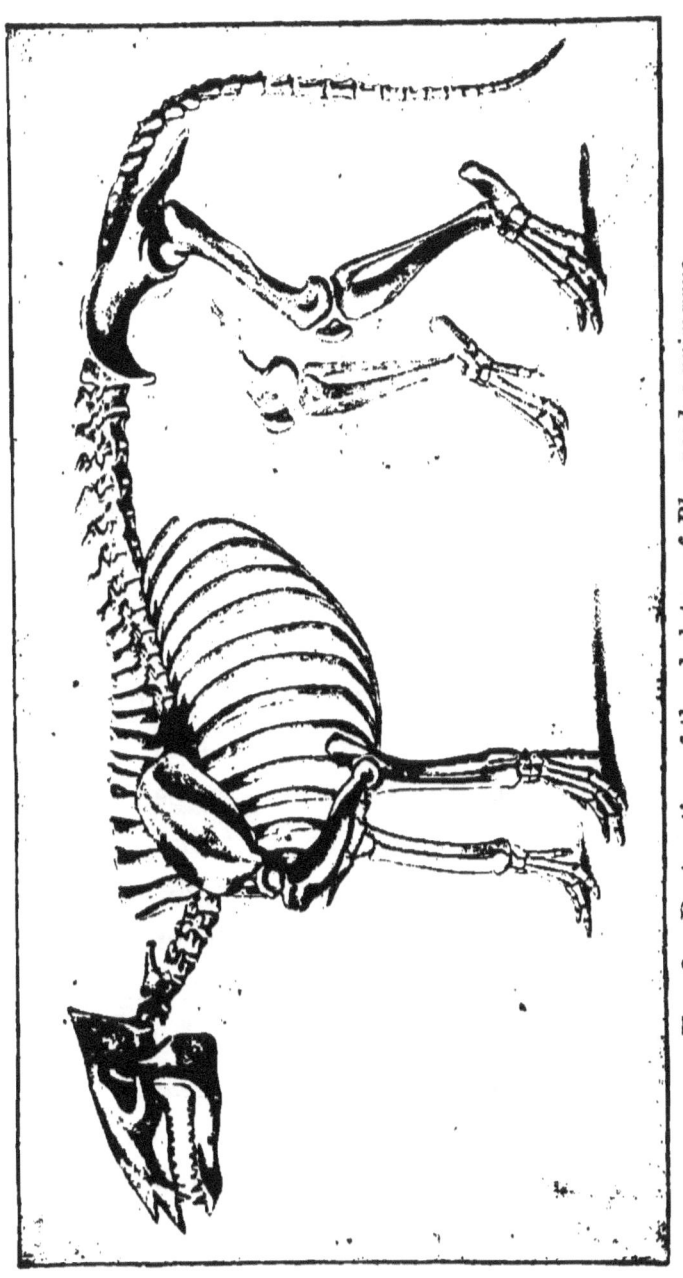

Fig. 2.—Restoration of the skeleton of *Phenacodus primævus.*
(*From the early Eocene of North America.*)

early Eocene formations in Switzerland. None of these have been found in such a complete state as Phenacodus; it is scarcely necessary therefore to dwell upon them here, though it will be well to devote a little time to the consideration of the structure of this form, which, if not in the actual ancestral line of all ungulates (and also, according to Cope, of the insectivores, carnivores, lemurs, monkeys, and even Man himself!) at all events exhibits the primitive pattern upon which the skeletons and teeth of all the others have been constructed, and which has never been departed from in any of them, however disguised by the special modifications of particular parts.

No part of the organization of an animal is so important in judging of its general position and characters, and at the same time so perishable, as the brain, and yet in consequence of this organ in all the higher vertebrates being accurately adapted in size and form to the bony case which contains it, we have been able to obtain a wonderfully perfect knowledge of, at all events, the rough anatomy of the brain in many animals which ceased to exist ages ago. A cast of the interior of the skull of Phenacodus, made and figured by Cope, shows a remarkably low type, both as to size and form, as compared with any modern ungulate of corresponding dimensions. The hemispheres of the cerebrum are especially small, compared with the cerebellum and olfactory lobes. This is exactly in accord with what is now generally known of the progressive diminution of the size of the brain in all groups of animals the further back we pass from the present time.

The teeth were of the same number as in the great majority of Eocene mammals—namely, three incisor or front teeth, one canine or corner tooth, and seven cheek teeth, consisting of four premolars and three molars,* or eleven altogether on each side of the upper, and the same number in the lower, jaw, making a total of forty-four. These numbers are briefly expressed in the formula

$$i.\frac{3}{3}\ c,\frac{1}{1},p\frac{4}{4},m\frac{3}{3}=\frac{11}{11}\times 2=44.$$

This is an extremely important formula to remember, as it is, as just mentioned, the one most commonly met with in mammals of the early Tertiary periods, and therefore the most generalized condition of dentition among all the higher orders of the class, and the one from which, by suppression or loss of certain of the teeth, all the special modifications now

---

* The back teeth, grinding teeth, or cheek teeth are divided into premolars and molars or true molars. In the ungulates there are three or four of the former and always three of the latter, which are the hindermost of the series and not preceded by milk teeth.

3

met with have been derived.* The characters of the teeth, as well as their number, are of a generalized type. The incisors are small, subequal, and with cutting edges, and are set in a semicircular line. The canines are, however, distinctly differentiated from the other teeth, isolated from the incisors in front and from the premolars behind, and are moderately strong, conical, and pointed. The premolars and molars are in a contiguous series, and the former are distinctly defined from the latter by their simpler structure. Their crowns are all extremely brachydont, or short from above downwards, a character met with in all primitive forms. The true molars belong to the simplest, or "bunodont" † type, having four principal rounded cusps on the grinding surface of each, with smaller cusps between, making six altogether. (See Fig. 3, A, page 31.)

The head is of small size compared with the body generally. The orbits, or cavities for the eyes, are not completed by bone behind, but are widely continuous with the temporal fossæ on the side of the skull, as in all primitive forms. The vertebral

---

* In the pig and a few insectivora alone among existing mammals is this number retained. In all others the total number of teeth falls short of forty-four, although, as we shall see, some horses still retain (as an exceptional condition) the primitive formula.

† From the Greek *bounos*, a hill or mound.

column is said to consist of seven cervical, fourteen or fifteen thoracic, six or seven lumbar, and three to five sacral vertebræ. The tail is long and tapering, much longer than in any existing ungulate, as it must have reached quite to the ground in walking. The scapula or shoulder-blade has a very oval form, resembling that of a carnivore more than that of any existing ungulate. The clavicles or collar-bones, as previously mentioned, are lost.*

* In using this expression the assumption is made that Phenacodus, and, in fact, all other mammals, are derived from forms having clavicles, and that the absence of these bones is a case of specialization, an assumption supported by the facts that the presence of clavicles is the rule in birds, reptiles, and amphibia; that they are well developed in various orders of mammals not otherwise closely associated, as marsupials, edentates, insectivores, and primates; and that they are also found, though often in a more or less vestigial condition, in rodents and carnivores. These facts all tend to show, if they do not conclusively prove, that the presence of the clavicle is the typical condition, notwithstanding its complete absence in extensive groups of mammals, as the ungulates and cetacea.

Since the above was in type the discovery has been announced of the presence of a rudimentary and transient clavicle in an early embryo of a sheep. This affords a complete confirmation of the view above expressed, and is a most astonishing instance of the persistence of a structure in the embryonic condition, which has, as far as our evidence tells, been absent in the adults during the whole of the Tertiary period. H. Wincza, "Ueber ein transitorisches Rudiment einer knöchernen Clavicula bei Embryonen eines Ungulaten." *Morphol. Jahrbuch,* xvi. p. 647. 1890.

The humerus, or upper-arm bone, presents a character not found in any of the existing ungulates, although common in the carnivora—that of a perforation, or foramen above the inner condyle. The two bones of the forearm, are, as in all generalized forms, both fully developed, the ulna being of large size throughout its length. The structure of the wrist, or carpus, is of great interest from its extremely primitive condition, each bone of the second row standing directly beneath, and articulating almost entirely and only with, the corresponding bone of the first row.

The five digits, with their typical number of phalanges, are completely developed, the third being the longest and strongest. The terminal or ungual phalanges are expanded, flattened, and rather spatulate, and evidently bore hoofs rather than claws. Each digit has a metacarpal bone and three phalanges, except the first or pollex (corresponding to the thumb of man), which has but two.

In the hind leg, the femur or thigh-bone shows considerable evidence of the presence of that projecting ridge on the outside, known as the third trochanter, found in all Perissodactyles, but in none of the Artiodactyle section. The two bones of the lower leg, the tibia and fibula, are distinct and complete.

In the ankle or tarsus the cuboid articulates with the calcaneum only. The astragalus presents a uniformly convex distal articular surface, as in Carnivora, but it has a trochlea or pulley-like proximal end, which the still more generalized *Periptychus* has not. The toes are five in number, much resembling those of the fore-foot. The animal was apparently not plantigrade, or walking with the whole of the sole of the foot, from the heel or hock to the toes, on the ground, as the bears do, nor did it walk on the tips of the toes only, as the horse does, but probably habitually stood in an intermediate position, with the heel raised more or less from the ground.

The remains of animals referable to this genus already discovered in the Wasatch Eocene are remarkably numerous, and differences in size and details of conformation have enabled Cope to describe and name nine species considered to be distinct from each other. They vary in size from that of a bulldog to a leopard or sheep. The structure of the bones of the nasal region has led to the suggestion that the head may have carried a short proboscis like that of the tapir.

As mentioned above, Phenacodus is not an isolated form, and allied but less perfectly known species appear to bridge over the interval between it and the next that will be spoken of.

In the year 1839, Sir Richard Owen described an imperfect skull of a small animal, not larger than a fox, which was discovered in the London Clay (Lower Eocene) of Herne Bay, in Kent, under the name of *Hyracotherium*, a name implying a supposed affinity (which we now know it does not possess) to the existing genus, *Hyrax*.* Specimens of identical or similar forms were subsequently found in Eocene formations in England and other parts of Europe, and others referred to the same genus far more abundantly and in a far more perfect state of preservation in beds of corresponding age in North America. To a closely allied form the name of *Pachynolophus* has been given, while *Pliolophus* and *Orohippus* are probably identical, and they are all so nearly related to a previously known but larger animal, called by Cuvier *Lophiodon*, that they are commonly associated to

---

* "Hyrax" (a Greek word for an animal which cannot be identified with certainty, perhaps a kind of shrew) is a name given by modern zoölogists to a small group, consisting of about a dozen species, of animals inhabiting the rocky districts of Syria and various parts of Africa, and which are of such peculiar structure that they are completely separated from all the existing and all the hitherto discovered extinct forms of life. They form the order *Hyracoidea* of Huxley, but may be included in the Ungulata, using that term in the very widest sense. They are the animals whose Hebrew name is translated in the English Bible into "coney" or rabbit, to which in size, color, and habit they bear a considerable general resemblance.

form a family *Lophiodontidæ*. Some of these animals* present a very distinct advance in evolution upon Phenacodus, an advance in some respects so great as to move them, according to Cope, into a distinct ordinal division of the Mammalia. This consists mainly in a modification of the form and relations of the bones of the carpus and the tarsus from their primitive condition, which modification persists in all the more recent forms of true ungulates, both Perissodactyle and Artiodactyle, and the absence of which is the principal distinction between them and the *Proboscidea* (elephants) and the *Hyracoidea*.

The bones of the second row of the carpus no longer stand exactly below the corresponding bones of the first row, but are all shifted a little way to the inner side of the foot, a change which is facilitated by the disappearance of the first digit, and which, with certain alterations in the form of the articular surfaces, tends to produce a more perfect interlocking of these bones one with another, and

---

* Madame Marie Pavlow has shown that under the name of *Hyracotherium* some very different forms have been confounded, the type species of Owen being the most primitive, and perhaps identical with Cope's *Phenacodus*, while the American *H. venticolum*, of which the whole skeleton is known, and to which the description in the text chiefly applies, is a much more advanced form.

thereby greater stability to the carpal region as a whole (see Fig. 1, p. 15). A corresponding change in the tarsus brings the cuboid into articular relation with the astragalus, which it wants in the primitive condition.

In the number of the digits a considerable modification has taken place in both feet. In the fore-limbs, instead of five, there are but four toes, a number which was long retained by a considerable section of the order, and persists even to the present day in the one family of tapirs. A foot thus formed may appear at first sight to belong to the Artiodactyle or even-toed type, especially as the missing toe is the first, and the four that remain are exactly those of the Artiodactyles. But on examining a little more closely it will be seen to present all the structural characteristics of the five-toed Perissodactyles, only changed by the removal of the first toe. The third is still the largest, and forms the center of support; the second and third are of equal size and form a pair arranged on each side of it. The fifth is an odd toe with nothing to balance it on the inner side of the foot. There is no trace of the symmetry around a line drawn between the third and fourth toes, or of the equality of these two which is seen in Artiodactyles. By referring to the diagram at p. 15 (Fig. 1), it will be easily understood that A is not converted into B by merely taking away the digit I.

It can hardly be supposed that the change took place suddenly from a five-toed to a four-toed forefoot, and indications have been discovered of intermediate forms (*Eohippus* of Marsh) in which a rudimentary first toe, represented only by the metacarpal bone, existed, but these have not yet been fully described.

The hind foot of Hyracotherium presents a still greater modification, both the outer digits, first and fifth, having disappeared. The three middle toes,

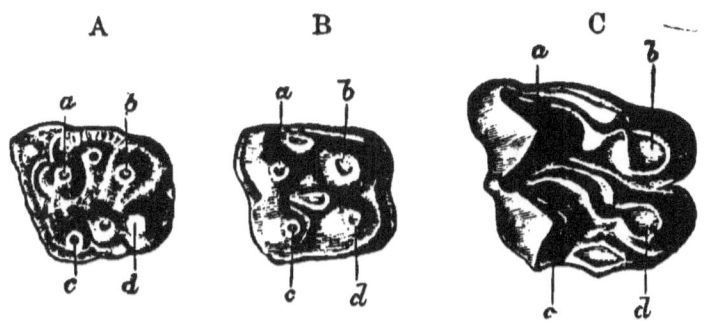

FIG. 3.—Grinding surface of upper molar tooth (very slightly worn). A, *Phenacodus*. B, *Hyracotherium*. C, *Anchitherium*. *a*, antero-external cusp; *b*, anterointernal cusp; *c*, postero-external cusp; *d*, postero-internal cusp.

symmetrically disposed to the axis of the third, are alone present. This is the condition of the hind foot of all known true Perissodactyles of the Eocene and Miocene epoch, and of the greater number of those still existing, the horse alone having advanced to a still more specialized state.

The teeth of Hyracotherium and its allies are the same in number as in Phenacodus—the same, in fact, as in the vast majority of Eocene mammals; but they begin to show more distinctly, especially in the grinding surfaces of the molars, a pattern in which the groundwork of all the subsequent complex modifications can be clearly traced (Fig. 3, B). The four larger corner cusps are distinct, but the intermediate ones are assuming the form of ridges or crests connecting the two anterior and the two posterior cusps respectively. These ridges are of a curved or sinuous form, and are not placed quite transversely, but have their inner ends inclining backwards. It will be useful to become thoroughly acquainted with this pattern, as it is the key to all others which will be hereafter spoken of.

In deposits of corresponding and more recent age to those in which the remains of Hyracotherium were found, immense numbers of bones and teeth have been discovered, indicating a variety of species more or less diverging in details, although constructed in the main upon the same type, the best known of which are included in the genera *Lophiodon* of European and *Hyrachyus* of American formations. Of the latter, remarkably complete skeletons have been discovered and fully described by Leidy, so that its osteology is now completely known. A few further stages of modification lead to the *Palæotherium* of the Paris basin (late Eocene), an interesting form from its association with the illustrious Cuvier, who in 1804 established its existence, and by comparison of its bones with those of all known recent species of animals, demonstrated for the first time to the satisfaction of the scientific world that animals had inhabited the earth other than those now found upon its surface. By this demonstration he laid the foundation of the study of palæontology of vertebrated animals—a study which has developed in this comparatively short period of time to such a marvelous extent, and which has still before it a future of unbounded promise.

By the time that the Palæotherium appeared, the group of Perissodactyles was already breaking up into different families by the gradual change in various directions from the primitive Lophiodont type. Some were passing step by step into tapirs, which still exist and retain much more of the original characters of the primitive ungulates of the Eocene period than any of the others now remaining on the earth, having indeed continued practically unchanged since the Miocene times; while almost all other mammalian forms which existed then have

either become extinct or undergone extensive modification.   In the structure of their feet they scarcely differ from Hyracotherium.   They are, in fact,
typical old conservatives, which have scarcely departed in any way from the manners, customs, or
structure of their ancestors.   They appear to be animals tending to extinction, for, though formerly
having a wide range of distribution through the continents of America, Europe, and Asia, they are now
only found at two rather isolated parts of the
world—*i.e.*, South and Central America and the
Malay region—and they are by no means numerous
either in species or individuals.

A second branch of the group can be traced
through such forms as *Hyracodon*, *Aceratherium*,
and *Aphelops* to the modern rhinoceroses, which
in many respects are more specialized than the
tapirs.   They have but three toes upon each foot,
and the teeth have been considerably changed,
some species having lost all the incisors or cutting
teeth of the front of the mouth.   They have, moreover, acquired the peculiarity of wearing one or
two large horns upon their noses, which the early
species of the family did not possess.

Somewhat allied to the rhinoceroses were some
remarkable animals which flourished in the early
Miocene time in North America, to various modifi-

cations of which the names *Menodus, Titanotherium, Megacerops, Brontotherium,* and *Symborodon* have been given.   They were of gigantic size, with a large head, having on the face a pair of stout, diverging, osseous protuberances, like the horn-cores of ruminants. Their fore-feet had four and their hind feet had three short stout toes. Also, out of the line of descent of any existing Perissodactyles was the remarkable *Macrauchenia,* a very specialized form, which existed in South America, apparently to Pliocene times, and then entirely disappeared from a world in which the conditions necessary for its well-being no longer existed, unless indeed we may suppose that the life-of a species, like that of an individual, comes to an end by virtue of some inherent tendency which is one of the essential attributes of its existence.   Leaving these and numerous other collateral branches which have left no representatives, we may pass to the third existing division, the most important in regard to the present subject.

Allied to Palæotherium, but probably more on the direct line of descent between Hyracotherium and the forms to be mentioned presently, was a small animal to which the name of *Palaplotherium* has been given, of which numerous teeth and bones have been found in the beds of Upper Eocene age at Hordwell in Hampshire, in the Isle of Wight, and in various parts of France and Germany. Another form associated by Cuvier with Palæotherium, the first known remains of which were found in the neighborhood of Orleans and hence called *P. aurelianense*, was by H. von Meyer separated generically under the name of *Anchitherium*. It flourished in the Miocene age, both in Europe and America, under many minor modifications, and is generally looked upon as in the direct line of ancestry of the modern *Equidæ*, which the true *Palæotherium* probably was not. One of the most striking characters by which it differs from Hyracotherium is the complete loss of the fifth digit of the fore-foot, all the extremities being alike in possessing only the three middle toes (second, third, and fourth of the typical condition), all reaching to the ground, but with the central one (the third) longer than the others (see Fig. 4). The two bones of the forearm (radius and ulna) and the two of the leg (tibia and fibula) were still quite distinct. The pattern of the grinding surface of the molar teeth (see Fig. 3, C, p. 31) had undergone some further modifications from that of Hyracotherium, which will be alluded to later on when describing the dentition of the horse. The Anchitherium was succeeded in the Pliocene period, in America, Europe, and Asia, by animals which have been named *Hipparion, Hippotherium, Protohippus*, and *Pliohippus*, of which there were many kinds, differing slightly in form and proportions, and in the characters of the

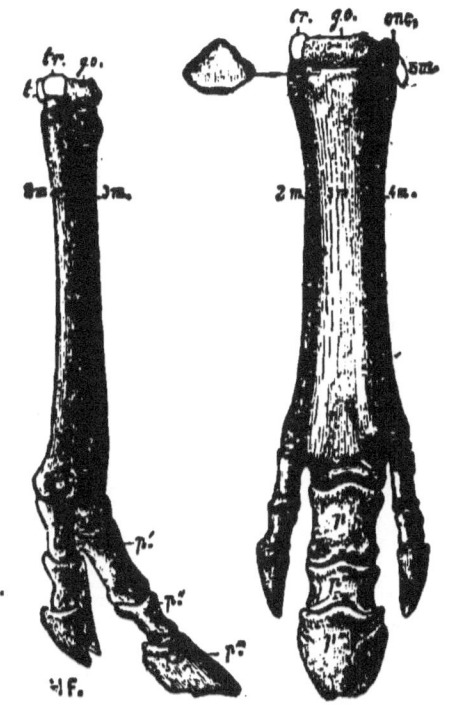

FIG. 4.—Side and front view of the bones of the left fore-foot of *Anchitherium* (without the upper row of carpal bones). *t*, trapezium; *tr*, trapezoid; *go*, magnum; *onc*, unciform; *2m*, second metacarpal; *3m*, third metacarpal; *4m*, fourth metacarpal; *5m*, rudiment of the fifth metacarpal; p', p'', p''', the first, second, and third phalanges of the middle (third) digit. The upper surface of third metacarpal is represented for comparison with Figs. 5 and 6, showing gradual change of form. (From Gaudry.)

enamel foldings of the molar teeth, but resembling each other in the structure of the feet. The lateral toes (second and fourth), though containing the full number of bones, were much reduced in size, and did not reach to the ground (see Fig. 5), but were suspended to the outside of, and rather behind, the large middle one, like the rudimentary outer toes of the

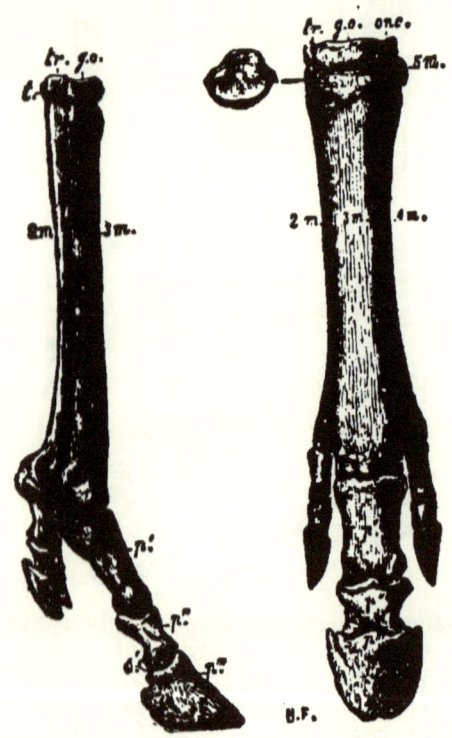

Fig. 5.—Side and front view of the bones of the left fore-foot of *Hipparion.* s, and s', upper and lower sesamoid bones. (From Gaudry.)

deer or the short first digit ("dew-claw") of the dog. Well-preserved remains of animals with this structure of foot have been met with abundantly at Pikermi, in Greece, and also in most of the deposits of corresponding age in the southern parts of France and Germany. Their former existence in England

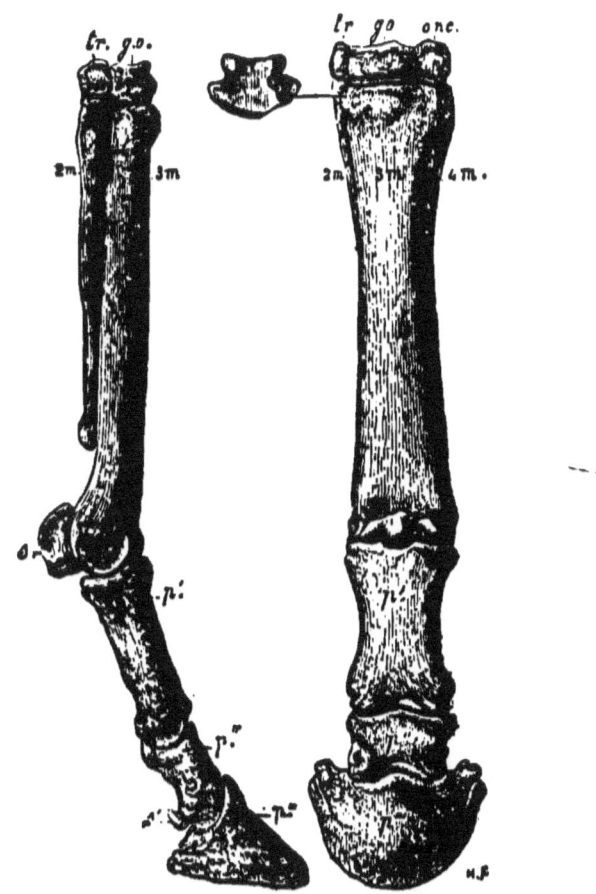

FIG. 6.—Side and front view of the bones of the left fore-foot of the horse. Letters as in Figs. 4 and 5. (From Gaudry.)

is only attested by scanty fragments found in the red crag of Suffolk. Horses, or rather horse-like creatures, with this structure of feet were no longer

4

met with when the Pleistocene, or latest geological period, set in; but then, for the first time, appeared the true horse, in its development exactly or very nearly as we know it now. The outer toes (second and fourth) were reduced to rudiments of the meta-carpals or metatarsals entirely concealed beneath the skin, while the middle or third toe was greatly elongated and had its last bone or ungual phalanx much expanded in breadth (Fig. 6). At the same time, the stability of the forearm and leg was increased by the two bones contained in each limb in the primitive forms becoming completely fused into one. Even since the Pleistocene period a change has taken place, as in horses of the present time the lateral rudimentary metapodials, or "splint bones" of veterinary anatomy (Fig. 6, 2m and 4m), though independent bones in the young animal have a great tendency to become united with the large mid-dle bone as life goes on; but in horses of the pre-historic or still earlier periods they are always found free, and were also relatively longer than they are now.

These modifications of the limbs thus gradually acquired in the course of time must have been asso-ciated with gradually increased speed in running, especially over firm and unyielding ground. Short, stout legs and broad feet, with numerous toes, spreading apart from each other when the weight of the creature is borne on them, are sufficiently well adapted for plodding deliberately over marshy and yielding surfaces, and the tapir and the rhinoceros, which in the structure of the limbs have altered but little from the primitive Eocene forms, still haunt the borders of streams and lakes and the shady depths of forests, as was probably the habit of their ancient representatives; while the horses are all inhabitants of the open plains, for life upon which their whole organization is in the most eminent degree adapted. The length and mobility of the neck, position of the eye and ear, and great development of the organ of smell, give them ample means of becoming aware of the approach of enemies; while the length of their limbs, the angles the different segments form with each other, and especially the combination of firmness, stability, and lightness in the reduction of all the toes to a single one, upon which the whole weight of the body and all the muscular power are concentrated, give them speed and endurance surpassing that of almost any other animal.

Remarkable changes in the structure and mode of growth of the teeth, which will be described in detail later on, have taken place *pari passu* with the modifications of the limbs and added greatly to their power as organs of mastication, and enabled their possessors to find their sustenance among the comparatively dry and harsh herbage of the open plains, instead of being limited to the more succulent vegetable productions of the marshes and forests in which their predecessors mainly dwelt.

The structural transitions from the diminutive Hyracotherium of the early Eocene period to the modern horse have been accompanied by a gradual increase of dimensions. The Miocene Anchitherium was of the size of a sheep. The Pliocene Hipparion and its allies were as large as modern donkeys; and it is only in the Pleistocene period that *Equidæ* appeared that approached in size the existing horse, the largest races of which are all the products of good feeding and selective breeding since it has become a domesticated animal.

It will be seen from what has been already said that the history of the Perissodactyles as a group offers many points of interest to the naturalist. Among these are its rapid extension and separation into various modifications, all containing numerous minor variations; the complete extinction of many of these, and the survival of three branches only, all of which (except the two domesticated species of the equine branch, which have been largely multiplied and diffused by man's agency) are poor in genera and species and far more restricted than formerly in geographical distribution. When we consider how extremely imperfect our knowledge of the former inhabitants of the earth must necessarily be, compared with that of those now existing, it is remarkable that we have already evidence enough to show that, at any period we may select since the Middle Eocene time, Perissodactyles were far more abundant, varied, and widely distributed than they are at present. This is the more interesting, as it is in marked contrast with what we know of the history of the other great division, the Artiodactyles, the latest modification of which, especially the hollow-horned ruminants or *Bovidæ* (antelopes, sheep, or oxen), are now the dominating members of the great Ungulate order, widespread in geographical range, rich in generic and specific variation, and numerous in individuals.

Of the three existing families of Perissodactyles, the least modified are, compared to their former abundance, in the most decadent state, while the most recently formed, and greatly modified, and most progressive group was until very recently bravely holding its own, in at least one region of its former extensive range. On the great plains of the African continent, zebras and quaggas roamed in countless herds within the memory of living man, and, except for his interference, there seemed no reason why they might not have continued to do so for ages yet. Explorers, hunters, and settlers, accompanied by the introduction of fire-arms into their native haunts, have, however, settled their doom. If events proceed as they are now doing, we may safely predict that the time is not very far distant when any living animal of the entire group of Perissodactyles, except in a state of domestication, will be a thing of the past. Under the rapidly changing circumstances of the world, caused by the spread of European civilization, it is not unlikely that the most ancient form, the tapir, may yet survive all the others, simply because it offers less inducement for the exercise of the destructive propensities of the modern sportsman, who is more responsible than any one else for the change now taking place in the normal balance of animal life on the earth's surface. In the next chapter this part of the subject will be entered into rather more fully.

CHAPTER II.

The tapirs (Family *Tapiridæ*)—Characters, species, geographical and geological distribution—The rhinoceroses (Family *Rhinocerotidæ*)—The horses (Family *Equidæ*)—Their
immediate predecessors—The hipparions or three-toed
horses of Europe and America—Existing species of horses
—The horse (*Equus caballus*)—Wild, domesticated, and
feral horses—Wild asses—*Equus hemionus* of Asia and its
varieties—The African wild ass and the domestic ass
(*Equus asinus*)—Striped members of the equine family—
Zebras and quaggas (*Equus zebra, E. burchelli, E. grevyi,*
and *E. quagga*)—Hybrids or mules—Aptitude for domestication only found in certain members of the family.

As shown in the last chapter, the Perissodactyle
ungulates, by various and gradually progressing deviations from the common original type, began at a
very early age to break up into several groups, some
of which, after undergoing a considerable degree of
specialization, have become extinct, without leaving
successors; but three of these modified types, already
distinct at the close of the Eocene period, have continued up to the present day, gradually, as time advanced, becoming more and more divergent from each

other. These are now represented by the three families of the Tapirs, the Rhinoceroses, and the Horses. Great as may be the differences between these animals as we see them now, we can trace their history step by step, as shown by the fragments preserved from former ages, farther and farther back into time, their differences continually becoming less marked, and ultimately blending together, if not into one common ancestor, at all events into forms so closely alike in all essentials that no reasonable doubt can be held as to their common origin.

As already indicated, the first named, the tapirs, have retained much more of the original characters of the primitive ungulates of the Eocene period than either of the others, and have indeed remained practically unchanged since the Miocene period; while almost all other mammalian forms which existed then have either become extinct or undergone extensive modification.

The Tapirs. (Family *Tapiridæ*.)

The tapirs constitute the single genus *Tapirus*, of which all the known species are much alike in external as well as anatomical characters. They are rather heavy, thick-set animals, with short and stout limbs. The fore-feet have four distinct toes, the first (that corresponding to the thumb of man) only

being absent, those that are present corresponding
to the second, third, fourth, and fifth of the typical
five-toed limb.   The third toe is the longest, the
second and fourth nearly equal, and the fifth the
shortest, and scarcely reaching the ground in the ordinary standing position.   The hind feet have three

A   B

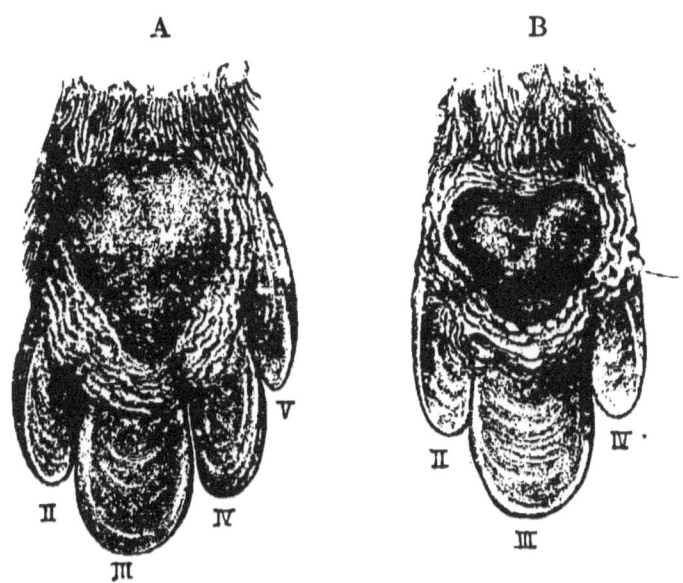

FIG. 7.—Plantar surface of right fore (A) and hind
(B) foot of Malay Tapir. (From Murie. *Journal
of Anatomy and Physiology*, vol. vi.)

toes, the middle one being the largest, and the two
others rather shorter.   Each toe is incased in a distinct, somewhat oval hoof, and there is a large,
roundish, callous pad on the sole of the foot, on
which the animal rests as well as on the toes (see

Fig. 7). The nose and upper lip are elongated into a flexible mobile snout or short proboscis, at the end of which the nostrils are situated. The eyes are rather small. The ears are of moderate size, ovate, and erect. The tail is very short. The skin is thick and smooth, and covered with a short and rather scanty hairy coat.

The skull is elevated and compressed. The orbit and temporal fossa are widely continuous, there being no true post-orbital process of the temporal bone. The anterior narial apertures are very large, and extend high on the face between the orbits. The nasal bones are short, elevated, triangular, and pointed in front. Vertebræ—Cervical, 7; dorsal, 18; lumbar, 5; sacral, 6; caudal, about 12. The teeth are—Incisors, $\frac{3}{3}$; canines, $\frac{1}{1}$; premolars, $\frac{4}{4}$; and molars, $\frac{3}{3}$ on each side, making a total of 42; thus, one tooth of the typical dentition, the first lower premolar, is wanting. The molar teeth of both jaws may be briefly described, as bilophodont, or having two transverse ridges; brachyodont, or short crowned, and without cement.

The species of tapirs are not numerous, and are all much alike in general form, anatomical structure, and especially dentition, so they may be said to constitute a single genus, *Tapirus*. The existing species may be grouped into two sections, the distinctive characters of which are only recognizable in the skeleton.

(A) With a great anterior prolongation of the ossification of the nasal septum (mesethmoid), extending in the adult far beyond the nasal bones, and supported and embraced at the base by ascending plates from the maxillæ. This section constitutes the genus *Elasmognathus* of Gill. There are two species, both from Central America.

Tapirus bairdi (Baird's tapir) : Mexico, Honduras, Nicaragua, Costa Rica, and Panama.

Tapirus dowi (Dow's tapir) : Guatemala, Nicaragua, and Costa Rica.

(B) With the ossification of the septum not extending farther forward than the nasal bones. Three species :

T. indicus (the Malay tapir) : Malay Peninsula (as far north as Tavoy and Mergui) and the islands of Sumatra and Borneo. *T. americanus* (*T. terrestris*, Linn.). The common South American tapir of the forests and lowlands of Brazil and Paraguay and the northern part of the Argentine Republic.

T. roulini. The Pinchaque or Roulin's tapir of the high regions of the Cordilleras of Colombia and Ecuador, 7,000 to 8,000 feet above the sea level.

The Malay tapir is the largest of the genus, and distinguished from all others by its peculiar coloration, the head, neck, fore and hind limbs being glossy black and the intermediate part of the body white. The demarkation of the two colors is distinctly de-

FIG. 8.—The American Tapir (*Tapirus americanus*).

From a photograph by Mr. Gambier Bolton of an animal living in the Zoölogical Gardens, London.

fined. The white of the body does not quite meet below, a median dark line intervening. All the American species are of a nearly uniform dark brown or blackish color when adult; but it is a curious circumstance that when young (and in this the Malay species agrees with the others) they are conspicuously marked with spots and longitudinal stripes of white or fawn color on a darker ground.

The habits of all the kinds of tapirs appear to be very similar. They are solitary, nocturnal, shy, and inoffensive, chiefly frequenting the depths of shady forests and the neighborhood of water, to which they frequently resort for the purpose of bathing, and in which they often take refuge when pursued. They feed on various vegetable substances, as shoots of trees and bushes, buds and leaves. They are hunted by the natives of the land in which they live for the sake of their hides and flesh.

"The tapirs," Wallace says, " form a small group of mammals, whose discontinuous distribution plainly indicates their approaching extinction." This view is supported, and the singular fact of the existence of so closely allied animals as the Malayan and the American tapirs in such distant regions of the earth and in no intervening places, is accounted for by what is known of the geological history of the race; for, if we may judge from the somewhat scanty remains which have been preserved to our times, consisting chiefly of teeth, the tapirs must once have had a very wide distribution. There is no proof of their having lived in the Eocene epoch; but by the Middle Miocene, tapirs undistinguishable generically from those now existing were already formed, so that they share the honors with *Hyomoschus* or *Dor-catherium* * of being the oldest living mammalian form. Such remains of Miocene and Pliocene age have been found in France, Germany, and England (Suffolk red crag). They appear, however, to have become extinct in Europe before the Pleistocene period, as none of their bones or teeth have been found in any of the caves or alluvial deposits in which those of elephants, rhinoceroses, and hippopotamuses occur in abundance; but in other regions their distribution at this age was wider than at present, as they are known to have extended eastward to China, and westward over the greater part of the southern United States of America, from South Carolina to California. Lund also distinguished two species or varieties from the caves of Brazil. Thus we have no difficulty in tracing the common origin of the now widely separated American or Asiatic species. It is, moreover, interesting to observe how very slight an amount of variation has taken place in forms isolated during such an enormous period of time. This may be owing to the extreme similarity of the conditions of existence in a Brazilian and a Malayan forest.

* A small Artiodactyle, somewhat intermediate in structure between a deer and a pig, found living in Western Africa and fossil in deposits of Miocene age in Germany.

THE RHINOCEROSES (Family *Rhinocerotidæ*).

The name *rhinoceros* (meaning in Greek "nose-horn") was applied by the ancients to an animal the most striking external peculiarity of which was certainly the horn growing above its nose.

The following are the general characters applicable to all the existing members of the family:

Head large. Ears of moderate size, oval, erect, prominent, placed near the occiput. Eyes small. Neck short. Skin very thick, in some species consisting of massive, indurated, almost inflexible, plates, with thin soft intervals or joints, to allow of motion. Hairy covering scanty. Tail of moderate length, slightly tufted. Limbs stout, rather short. Three completely developed toes, with distinct, broad, rounded hoofs on each foot.*

All existing species have one or two horns, placed in the middle line upon the face. When one is present, it is situated over the conjoined nasal bones; when two, the hinder one is over the frontals. These horns differ in details of structure from those of any other animal, though belonging to the same category of epidermic growths as the horns of oxen, as well as nails, claws, hoofs, callosities, and warts. Their

---

* In some extinct species a small outer toe is present on the fore-foot.

structure, as seen under the microscope, has a great resemblance to that of whalebone, being composed of a solid mass of hardened epidermic cells growing from a cluster of long dermal papillæ. The cells formed on each papilla constitute a distinct horny fiber, like a thick hair, and the whole are cemented together by an intermediate mass of cells which grow up from the interspaces between the papillæ. It results from this that the horn has the appearance of a mass of agglutinated hairs, which, in the newly-growing part at the base, readily fray out on destruction of the softer intermediate substance; but any one acquainted with the structure and mode of growth of true hairs will see that the fibers differ from them in growing around a long free papilla on the surface of the derm, instead of from a very short papilla sunk deeply in a follicular involution of the same. These horns are really warts, which have assumed a solid and definite form, and the stages by which they may have developed are illustrated in the irregularly-shaped supplementary horns which are sometimes found either on the face or other parts of the body, the product of some local abnormal condition of the skin.*

* See a case of an African rhinoceros with a third horn described in the *Proceedings of the Zoölogical Society of London*, 1889, p. 418.

When fully developed, the horns are of a more or less conical form usually curved backwards at their ends, and often grow to a great length (three or even four feet), but they are constantly worn away at the ends and sides by being rubbed against trees or stones, and are continually growing at the base. Their length and shape are, therefore, subject to considerable variation, even in the same individual at different times, and so cannot be depended upon for the distinction of species, as some naturalists have imagined. Though not normally shed, they are occasionally torn off at the base,* in which case a new horn will grow in its place, although, if the matrix, or portion of the skin to which it is attached, is much injured, it may assume a more or less irregular shape.

As regards the dentition, the incisors are variable, generally reduced in number, and often quite rudimentary and entirely disappearing at a very early age.† The canines in existing species are

---

* This happened in 1870 to the male Indian rhinoceros still living in the gardens of the Zoölogical Society of London, in an effort to raise with its horn a strong transverse iron bar at the lower part of the railings of the inclosure in which it was confined.

† It is difficult to see what advantage the great African two-horned rhinoceroses can find in the complete absence of their front teeth, but this is one of those numerous cases in which we must be content to acknowledge our ignorance and wait for the explanation.

absent.* In respect to the front teeth, therefore, a
very marked amount of specialization has taken
place. On the other hand, the cheek teeth are retained in full normal numbers—viz., four premolars
and three molars on each side above and below, all
in contact, and closely resembling each other, except
the first, which is much smaller than the rest, and
often lost early in life. The others gradually increase in size from before backwards up to the
penultimate, which is the largest. The upper molars have a very characteristic pattern, admirably
adapted for bruising and crushing coarse vegetable
substances, and which is clearly a modification of
the pattern already seen in the corresponding teeth
of *Hyracotherium.* The lower molars are of simpler
form, the two transverse ridges being curved into
a crescentic form. In neither case are the deep depressions between the ridges filled up with cement,
as in the horse.

The skull is elongated and elevated posteriorly
into a transverse occipital crest. It has retained its
primitive condition in possessing no post-orbital
processes or any separation between the orbits and
temporal fossæ. The nasal bones are large and
stout, co-ossified, and standing out freely above the

---

* It should be stated that certain teeth, regarded above as
incisors, are considered by some zoölogists as modified canines.

premaxillæ, from which they are separated by a
deep and wide fissure; the latter bones are very
small, generally not meeting in the middle line in
front, often quite rudimentary, a specialization con-
current with the loss of the upper incisor teeth.   The
brain cavity is very small for the size of the skull.
Vertebræ—Cervical, 7; dorsal, 19–20; lumbar, 3;
sacral, 4; caudal, about 22.

The *Rhinocerotidæ* are all animals of large size,
but of little intelligence, generally timid of disposition, though ferocious when attacked and brought
to bay, using the nasal horns as weapons, with which
they strike and toss their assailants.   Their sight is
dull, but their hearing and scent are remarkably
acute.   They feed on herbage, shrubs, and leaves
of trees, and, like so many large animals which inhabit hot countries, sleep the greater part of the day,
being most active in the cool of the evening or even
during the night.   They are fond of bathing or wal-
lowing in the mud.   None of the species have been
domesticated.   The family once contained many
more species and was much more widely distributed
than at present.   As already indicated, our knowledge of them is as yet but fragmentary, though constantly augmenting, especially by discoveries made
in the Tertiary deposits of North America, a region
from which they all died out long ago, though, judg-

ing from the evidence at present available, this was
the locality in which they first made their appearance.
In the Eocene formations of the Rocky Mountains
are found the remains of numerous modifications of
the primitive Perissodactyle type, from which the
rhinoceroses may have originated.   In the Lower
Miocene a form called *Hyracodon* by Leidy already
presented many of the characteristics of the family,
though, especially as regards the dentition, still in
a very generalized condition.   It had, however,
already lost the fifth toe of the fore-foot.   The next
stage of specialization is represented by *Aceratherium* and *Aphelops*, found in the Miocene of Europe
and America, which still, like the last, show no sign
of having possessed a nasal horn.   The former
differs from the existing species, and also from *Hyracodon*, in having four toes on the anterior limb
instead of only three.   At the same period forms
occurred (*Diceratherium*, Marsh) which show a pair
of lateral tubercles on the nasal bones apparently
supporting horns side by side.   These, however, soon
disappeared and gave way in the Old World to species
with one or two horns in the median line, a stage of
development which apparently was never reached in
America.   In the Pliocene and Pleistocene of Europe
and Asia numerous rhinoceros remains have been
found, all more or less nearly related to the existing

species. The present African two-horned type was already represented in the early Pliocene of Greece by *R. pachygnathus*, the skeleton of which is described by Gaudry as intermediate between the existing *R. bicornis* and *R. simus*. As many as three species were inhabitants of the British Isles, of which the best known is the Tichorhine or woolly rhinoceros, *R. antiquitatis* of Blumenbach, *R. tichorhinus* of other authors, nearly whole carcasses of which, with their thick woolly external covering, have been discovered, associated with those of the mammoth, preserved in the frozen soil of the north of Siberia, and which, in common with some other extinct species, had a solid median wall of bone supporting the nasals. From this peculiarity it has been inferred that the horns were of size and weight surpassing those of the modern species. The one-horned Indian type was well represented under several modificatons (*R. sivalensis, R. palæindicus*, etc.) in the Pliocene deposits of the sub-Himalayan region, and forms more allied to the African bicorn species have also been found in a fossil state in India. *R. schleiermacheri* of the late European Miocene in some features, especially in possession of incisor teeth and two horns, resembled the existing Sumatran rhinoceros, but it differed in important cranial characters.

The existing species of rhinoceros are naturally

grouped in three sections, which some zoölogists consider of generic value.

I. *Rhinoceros* proper. The adults with a single large compressed incisor tooth above on each side, and occasionally a very small lateral one; below, a very small median, and a very large, procumbent, pointed, lateral incisor (or canine?). Nasal bones pointed in front. A single nasal horn. Skin disposed in very massive, definitely arranged armor-like plates, with soft interspaces or joints between them.

There are two well-marked species of one-horned rhinoceros:

1. The Indian rhinoceros, *R. unicornis* of Linnæus,* the largest and best known, from being the most frequently exhibited alive in England, is at present only met with in a wild state in the Terai region of Nepal and Bhutan, and in the upper valley

---

* Many authors use Cuvier's name, *R. indicus*, in preference to this, on the ground that there are more than one species with one horn, forgetting that the name substituted is equally inconvenient, as more than one species live in India. The fact of a specific name being applicable to several members of a genus is no objection to its restriction to the first to which it was applied, otherwise changes in old and well-received names would constantly have to be made in consequence of new discoveries. Ill-considered attempts at precision of nomenclature are often sources of confusion and future difficulty. As Huxley has truly said, "It is better for science to accept a faulty name which has the merit of existence, than to burden it with a faultless newly-invented one."

of the Bramaputra or province of Assam, though it formerly had a wider range. The first rhinoceros seen alive in Europe since the time when they, in common with nearly all the large remarkable beasts of both Africa and Asia, were exhibited in the Roman

FIG. 9.—Indian Rhinoceros (*Rhinoceros unicornis*).

From a photograph by Mr. Gambier Bolton of an animal living in the Zoölogical Society's Gardens. In wild animals the horn often grows to a greater length.

shows, was of this species. It was sent from India to Emmanuel, King of Portugal, in 1513; and from a sketch of it taken in Lisbon, Albert Dürer composed his celebrated, but rather fanciful, engraving, which was reproduced in so many old books on natural history.

2. The Javan rhinoceros (*R. sonduicus*, Cuvier) is distinguished by smaller size, special characters of the skull and teeth, and different arrangement of the plications of the skin, especially in the deep depression which runs upwards and backwards from the middle of the side of the neck, passing over the back, joining its fellow on the opposite side, and thus isolating a plate proper to the neck from the great shoulder-plate. In the Indian rhinoceros (Fig. 9) this fold or depression does not pass over the back, but curves backwards and is lost above the shoulder. This species has a more extensive geographical range than the last, being found in the Bengal Sunderbuns near Calcutta, Burmah, the Malay Peninsula, Java, Sumatra, and probably Borneo. A hornless rhinoceros (*R. inermis*) which has been described is supposed to be the female of this species, but this is a point which requires further investigation.

II. *Ceratorhinus.* The adults with a moderately-sized compressed incisor above, and a laterally-placed pointed procumbent incisor below, which is sometimes lost in old animals. Nasal bones narrow and pointed anteriorly. A well-developed nasal horn and a small horn behind it, separated by a considerable interval. The skin thrown into folds, but these are not so strongly marked as in the former section. The smallest living member of the family, the Sumatran rhinoceros (*R. sumatrensis*, Cuv.), belongs to this group. Its geographical range is nearly the same as that of the Javan species, though not extending into Bengal; but it has been found in Assam, Chittagong, Burmah, the Malay Peninsula, Sumatra, and Borneo. It is possible that more than one species have been confounded under this designation, as two animals now living in the London Zoölogical Gardens present considerable differences of form and color.

III. *Atelodus.* In the adults, the incisors are quite rudimentary or entirely wanting. Nasal bones thick, rounded, and truncated in front. Two horns, both well developed and in close contact with each other. Skin thick but smooth, without any definite thickened plates or permanent folds.

The two well-marked species are peculiar to the African continent:

1. The common two-horned rhinoceros (*R. bicornis*, Linn.) is the smaller of the two, with a pointed, prehensile upper lip. It ranges through the wooded and watered districts of Africa, from Abyssinia in the north to the Cape Colony, but its numbers are yearly diminishing owing to the inroads of European civilization, and especially to the persecutions of English sportsmen. It feeds exclusively upon leaves and branches of bushes and small trees, and chiefly

frequents the sides of wood-clad rugged hills. Specimens in which the posterior horn has attained a length as great as, or greater than, the anterior horn have been separated under the name of *R. keitloa*, but, as already mentioned, the characters of these appendages are too variable to found specific distinctions upon. The two-horned African rhinoceros is far more rarely seen in menageries in Europe than either of the three Indian species, but one has lived in the gardens of the London Zoölogical Society since 1868. Excellent figures from life of this and the other species are published in the ninth volume of the Transactions of the Society.

2. Burchell's, or the square-mouthed rhinoceros (*R. simus*), sometimes called the white rhinoceros, though the color (dark slate) is not materially different from that of the last species, is the largest of the whole group, and differs from all the others in having a square, truncated upper lip, and a wide, shallow, spatulate form of the front end of the lower jaw. In conformity with the structure of the mouth, this species lives entirely by browsing on grass, and is therefore more partial to open countries or districts where there are broad grassy valleys between the tracts of bush. It is only known in the regions south of the Zambesi, and owing to the causes indicated above has of late years become extremely scarce;

indeed, the time of its complete extinction cannot
be far off, if indeed it has not already arrived. No
specimen of this species has ever been brought alive
to Europe, and very few examples are to be seen in
our museums. The flesh of both species of African
rhinoceroses is considered very good eating by the
natives of the countries in which they live, being,
according to Selous,* "something like beef, but yet
having a peculiar flavor of its own. The part in
greatest favor among hunters is the hump, which, if
cut off whole, and roasted, just as it is, in the skin, in
a hole dug in the ground, would be difficult to match
either for juiciness or flavor."

Before leaving the rhinoceroses, a huge creature
belonging to the family, to which the name of *Elasmotherium* has been given, should be mentioned. It
is only imperfectly known by fossil remains found in
Pleistocene deposits in Russia; but it is interesting
on account of the remarkable degree of specialization
its molar teeth had attained, far beyond that of the
existing rhinoceroses, and comparable in the length
of the crowns and the complex folding of the enamel
to that of the horses of the same or later period,
though on a very much larger scale. It affords a
good illustration of the fact previously mentioned,

* See F. C. Selous, *Proceedings of the Zoölogical Society of
London*, 1881.

that the most highly specialized members of a group are not always those that survive the longest.

The Horses. (Family *Equidæ*.)

As has been already stated, at about the time of the world's history when the Miocene was passing into what we term the Pliocene epoch, there were no true horses in exactly the sense in which we use the word now, but horse-like animals were extremely abundant both in America and the Old World, differing from existing horses in details of teeth and skeleton, especially in the presence of three toes upon each foot, a large middle toe and a smaller one, not reaching to the ground, placed on each side of it. To these animals, the step from the *Anchitherium* of the early Miocene, mentioned in the last chapter, was not a very great one.

Unfortunately, when remains of this type were first discovered, two generic names were given to them almost simultaneously—*Hipparion* and *Hippotherium*, the former being a diminutive of *hippos*, the Greek for "horse"; the latter a compound of *hippos* and *therion*, a wild beast, Latinized to *therium*, a termination very commonly employed in modern scientific language when coining new appellations for extinct animals. The first name was given

by the French palæontologist Christol; * the latter
by Kaup of Darmstadt.† Although Christol's appears to have the actual priority, and has been ex-
tensively used, especially in France and England, it
does not seem to have been accompanied when first
brought out by any clear description, and is therefore
not acknowledged by many zoölogical authors, especially in Germany and America, where *Hippotherium*
takes its place.   *Protohippus* and various other names
have been proposed for other modifications (differing
chiefly in tooth structure) of animals in the same general phase of evolution.   The great variety of these
forms may be gathered from the fact that in a recent
memoir Professor Cope has described fifteen species of *Hippotherium*, which he considers to be quite
distinct from each other, from North America alone.‡

The term "Hipparion" has become so well known,
even beyond the limits of strictly scientific literature,
that it may be conveniently used as a common name
for all the three-toed horse-like animals which immediately precede the existing *Eqidæ*, reserving
Hippotherium, *Protohippus*, etc., for generic modifications capable of exact zoölogical definition.

---

* *Ann. Sci. Indust. Mid. France*, vol. i. p. 180 (1832).
† *Jahrbuch für Mineralogie*, etc., 1833, p. 327.
‡ "A Review of the North American Species of Hippothe-
rium," *Proc. American Philosophical Society*, 1889.

In the quarries of Pikermi, in Greece, an immense
number of remains of large animals, now entirely
extinct, have been discovered and made known to us
mainly by the admirable memoir published upon
them by the eminent French palæontologist, Albert
Gaudry.* These animals include monkeys, civets,
hyenas, wild boars, rhinoceroses, antelopes of various
kinds, a great giraffe-like creature called *Hellado-
therium*, and hipparions in such multitudes as to
show that these animals must have wandered over
the plains of Europe in great herds, comparable to
those of the wild asses of Tartary and the zebras of
South Africa of recent times. The collection made
by Gaudry alone consisted of 1,900 bones, belonging
to at least twenty-four individuals. They have also
been found in similar numbers at Eppelsheim in
Germany, and at Mont Léberon and in Vaucluse in
the south of France.

One of the principal characteristics of the skeleton
of the Pikermi hipparion is the presence of a considerable depression or pit upon the side of the face
in front of the orbit or cavity for the eye. Although
such a pit is not found in any of the existing species
of horse, it was not infrequent in many extinct
forms, and varied in them in size and depth. It so
closely resembles a similar depression, found in the

* *Animaux fossiles et Géologie de l'Attique*, 1862.

same situation in many species of deer and antelopes, which lodges a glandular infolding or pouch of the skin called the "suborbital gland," "crumen," or in French "larmier," that there can be little doubt but that it had the same purpose in the hipparion.   The gland in the existing animals that possess it secretes a peculiar oily, odorous substance, the scent of which enables the animals provided with it to recognize each other even at immense distances, the faculty of smell being also developed to a wonderful degree. At certain seasons of the year the glands are especially active, and their position is such that when the animal is feeding particles of the odorous secretions will fall on and adhere to the herbage around, and thus afford indications to any other animals of the same species that may for some time afterwards pass over the same ground.

The presence of this gland in the hipparion and its absence in the more modern *Equidæ* has been given as a reason for supposing that the latter are not the direct descendants of the former, but must have been derived from some other form in which such a specialization had not been developed.   This, of course, is probable; but it must not be forgotten that very slight changes in habits, or the increased power and use of other senses than that of smell, may have diminished the value of the information

afforded by means of this gland, and ultimately led
to the elimination of the organ itself. It may be
that a change from a life habitually passed in forests
or scrub to one in open plains would be sufficient to
account for such a modification in structure. In
any case, it is one which must be very easily brought
about, without any other great changes, as the modern ruminants show, being present or absent in them
quite irrespectively of real affinity, as indicated by
more fundamental and less superficial and adaptive
structural characters. It would be interesting to make
a careful microscopical examination of the skin of
this region in all existing species of *Equidæ*, to ascertain whether any traces of the gland can be found;
for it is present in a most rudimentary condition,
without showing any impression on the surface of the
bone below in several of the existing *Bovidæ*, the
sheep, for instance. In this animal, its place in the
economy of life is supplied by the curious little bot-
tle-like glandular pouches placed between the toes.

Another easily-recognized distinction between
the hipparion and all modern horses is seen in the
structure of the upper molar teeth. The anterior
inner cusp of the primitive form (Fig. 10, *a.i.*) constitutes a distinct column instead of being, as in the
horse, united for its whole length with the rest of the
tooth. The foldings of the enamel are also devel-

oped to a remarkable extent of complexity. These
characters cannot be clearly understood until the
details of the structure of the teeth, to be explained
in the next chapter, are known; but they are sufficient to enable any one conversant with them to
recognize a single molar of an hipparion from that
of any of the existing species, and to show that the

FIG. 10.—Section of upper molar tooth of hipparion,
from the Red Crag of Suffolk. *a.i*, anterior internal column completely isolated from the main mass
of dentine; *p.i*, posterior internal column. The
uncolored portion is the dentine, the shaded part
the cement, and the black line separating these
two the enamel. Compare with Fig. 21, *c*, p. 125.

horse-like teeth found occasionally among the *débris*
of former Miocene or Pliocene formations in the
Red Crag of Suffolk belong to animals of this group.

These dental characters, and also details in the
structure of the bones of the feet, have led even
more conclusively than the presence of the suborbital
depression to the view that the hipparion, or, at all
events, the European *Hippotherium gracile*, was not

6

on the direct line of descent of the modern horses, but that it was a form which, having attained a considerable degree of specialization in some particulars, a wide geographical distribution and great abundance of individuals, became, as has so often happened in similar cases, extinct without direct descendants from causes which we at present cannot divine. Perhaps an inability to lose the useless outer toes may have given it a disadvantage in a severe competition for existence with otherwise closely allied forms, which had already adopted the style of foot which clearly shows itself the best for the existing requirements of the race.

Judging from tooth-structure alone, a very perfect series of modifications from *Anchitherium* to the modern horses can be shown through various species of the American genera called *Merychippus* and *Protohippus*, without the intervention of the special characteristics of hipparion; but, unfortunately, of many of these forms, the bones, and especially those of the limbs, are known very imperfectly or not at all. There is, however, already enough to show that it is by no means impossible that America may have been the cradle of all the existing *Equidæ*, as it seems to have been of such apparently typical Old World forms as rhinoceroses and camels, and that they spread westward by means of the former free com-

munication between the two continents in the neighborhood of Behring's Straits, and, having prevailed
over the allied forms they found in possession, totally
disappeared from the country of their birth until
re-introduced by the agency of man.   This supposi-
tion, based upon the great abundance and variety
of the possible ancestral forms of the horse which
have lately been discovered in America, may be at
any time negatived by similar discoveries in the Old
World, the absence of which at the present time cannot be taken as any evidence of their non-existence.

In a popular exposition of this subject it would
be out of place to give an account of the views, more
or less crude, which have been put forth by the various zoölogists who have lately exercised much labor,
patience, and thought in endeavoring to investigate
the exact lines of descent of the different species and
even breeds of the existing horses from those of
earlier periods.   In the first place, they would only
be intelligible to any one possessing a full knowledge
of the minute anatomical characters on the comparison of which the results are based; but what is of
still more consideration, the conclusions from all
these researches can be looked upon at present as
provisional only, being founded upon such imperfect
materials as exist as yet in our collections, and liable
to be modified at any moment by fresh discoveries

which may be expected to be made from time to time.

That the science of palæontology has a great future before it has already been intimated. The recesses of the earth still teem with riches in untold numbers. When they have been brought to the light of day, their geological antiquity and their anatomical characters will offer a fruitful field for investigation and speculation. The harvest is indeed abundant and the laborers hitherto few. The excellent work done in this subject by Marie Pavlow, of Moscow, is therefore particularly interesting, as showing for the first time in the history of this branch of science that women are equally competent with men to enter into the field and join in gathering the golden grains of knowledge.*

* The following are some of the principal works from which fuller information concerning the palæontology of the *Equidæ* can be obtained :

E. Cope: "The Perissodactyla" (*American Naturalist*, Nov. 1887, p. 985). Numerous other memoirs by the same author in American scientific periodicals.

A. Ecker: "Das Europäische Wildpferd und dessen Beziehungen zum domesticirten Pferde" (*Globus*, Band xxxiv. 1878).

A. Gaudry: *Ancêtres de nos Animaux*, 1888; *Les Enchaînements du Monde animal*, 1878; *Animaux fossiles du Mont-Léberon*, 1873; *Géologie de l'Attique*, 1862–1867.

W. Kowalevski: "Sur l'*Anchitherium aurelianense* et sur l'histoire paléontologique des Chevaux" (*Mém. de l'Acad. Impér. de St.-Pétersbourg*, 1873).

Existing Species of Equidæ.

The members of the family *Equidæ* existing at the present time upon the earth are generally considered to belong to one genus, that designated *Equus* by Linnæus. As, however, a genus is a merely artificial assemblage of allied animals established for the convenience of nomenclature, zoölogists differ greatly among themselves as to the limits that

J. Leidy: *Extinct Vertebrate Fauna of the Western Territories*, and other memoirs.

R. Lydekker: Various memoirs and *Catalogue of Fossil Mammalia in British Museum*, Part III., 1886.

Forsyth-Major: "Beiträge zur Geschichte der fossilen Pferde," 1877 (*Schweizer paläontol. Gesellschaft*, vol. iv.).

O. C. Marsh: "Fossil Horses in America" (*American Naturalist*, 1874), and other memoirs.

A. Nehring: "Fossile Pferde aus Deutschen Diluvial-Ablagerungen" (*Landswirthschaftl. Jahrbuch*, 1884, Bd. xiii. Heft 1, p. 81).

Marie Pavlow: *Études sur l'histoire paléontologique des Ongulés.* Moscou, i. 1887; ii. 1888; iii. 1890.

L. Rütimeyer: "Beiträge zur Kenntniss der fossilen Pferde," 1863; "Weitere Beiträge zur Beurtheilung der Pferde der Quaternär-Epoch" (*Abhand. Schweizerischen paläont. Gesellsch.*, 1875).

M. Schlosser: "Beiträge zur Kenntniss der Stammesgeschichte der Hufthiere, und Versuch einer Systematik der Paar und Unpaarhufe" (*Morpholog. Jahrb.*, 1886, Bd. xii., Heft 1).

M. Wilckens: "Forschungen auf dem Gebiete der Paläontologie der Hausthiere" (*Biol. Centralblat.*, 1889).

J. L. Wortman: "On the Origin and Development of the Existing Horses" (*Kansas City Review of Science*, 1882, Nos. 2 and 12).

should be assigned to such a group, and there is a considerable tendency to break up the old and larger genera into smaller ones, if any characters can be found by which certain of the species can be associated together and distinguished from the others. In this way, the genus *Equus* has been separated into *Equus* proper, *Asinus*, and *Hippotigris*, the former containing the horse alone, the second the asses, and the third the zebras. The great inconvenience of altering the limits of genera is that, as the name of the genus is part of the name by which (in the prevailing binomial system of zoölogical nomenclature) the animal is designated in scientific works in all languages, every change in the limits of a genus involves some of those endless changes in names which are among the greatest causes of embarrassment in the study of zoölogy in modern times, and do so much to repel beginners from entering upon it.*

Although it may be convenient to recognize that the horse has special characters by which it is distinguished from the rest of the group, and that the others are all more nearly allied to each other than they are to it, and that the zebras, though otherwise

* The name of the genus, it must be remembered, in the binomial system corresponds to the surname or family name of persons of civilized nations, but in zoölogy it always precedes the specific name, which corresponds to our prename or Christian name.

closely related to the asses, are distinguished from them and associated together by their style of coloring and geographical distribution, it scarcely seems desirable that such distinctions should be made the ground of difference of generic appellation, and they will in this work all be spoken of as members of the genus *Equus*.

THE HORSE (*Equus caballus*, Linn.) is distinguished from all the others by the long hairs of the tail being more abundant and growing quite from the base as well as the end and sides, and also by possessing a small bare callosity on the inner side of the hind leg, just below the "hock" or heel-joint, in addition to the one on the inner side of the fore-arm, above the wrist or "knee," common to all the genus. The mane is also longer and more flowing, the front part of it drooping over the forehead, constituting the "forelock"; and the ears are shorter, the limbs longer, the feet broader, and the head smaller.

Though the existing horses are usually not marked in any definite manner, or only irregularly dappled (*i.e.* marked with large light spots surrounded by a darker ring), many examples are met with showing a dark streak running along the center of the back, like that found in all other members of the genus, and even with dark stripes on the shoulder and legs.

Darwin * collected a number of cases of horses of various breeds and countries so marked, and from them came to the conclusion of the "probability of the descent of all the existing races from a single dun-colored, more or less striped primitive stock, to which our horses occasionally revert."

Fossil remains of true horses, differing but very slightly from those now existing, are found abundantly in the most recent geological ages in almost every part of America, from Escholtz Bay in the north to Patagonia in the south. Whether any of these remains should really be referred to *E. caballus* or not—that is, whether they belonged to animals which possessed all the external characters attributed above to that species—is, of course, doubtful. Our knowledge of existing forms teaches us that closely similar and perhaps identical skeletal and dental characters may be associated with considerable external differences, especially in the character, distribution, and color of the hair. If zebras were only known from such portions of their structure as could be preserved in a fossil state, we should never have guessed how greatly they differed in outward aspect from horses and asses. All that we can do with a fossil bone or tooth is to assign it to any known species which it re-

* *The Variation of Animals and Plants under Domestication,* vol. i. chap. ii. (1868).

sembles so closely that no actual, definable difference between them can be detected. In this sense we may speak of *Equus caballus* having existed in America before its introduction by the Spaniards, although it is commonly supposed that at the time of the conquest no horses, either wild or domesticated, were to be found on the continent.* This is the more remarkable as, when imported from Europe, the horses that ran wild proved by their rapid multiplication in the plains of South America and Texas that the climate, food, and other circumstances were highly favorable to their existence. The former great abundance of *Equidæ* in America, their extinction, and their perfect acclimatization when reintroduced by man, form curious but as yet unsolved problems in geographical distribution.

In Europe, wild horses were extremely abundant in the Neolithic, or polished-stone period. Judging by the quantity of their remains found associated with those of the men of that time, the chase of these

* The usual statement as to the complete extinction of the horse in America is thus qualified, as there is a possibility of the animal having still existed, in a wild state, in some parts of the continent remote from that which was first visited by the Spaniards, where they were certainly unknown. It has been suggested that the horses which were found by Cabot in La Plata in 1530 cannot have been introduced. See M. Wilckens's "Forschungen auf dem Gebiete der Paläontologie der Hausthiere" (*Biolog. Centralblat.*, 1889).

animals must have been one of their chief occupations, and they must have furnished one of their most important food-supplies. The characters of the bones preserved, and certain rude but graphic representations carved on bones or reindeer's antlers found in several caves in the south of France,—enable us to know that they were rather small in size and heavy in build, with large heads and rough, shaggy manes and tails—much like, in fact, the present wild horses of the steppes of the south of Russia. These horses were domesticated by the inhabitants of Europe before the dawn of history. Cæsar found the Ancient Britons and the Germans using war-chariots drawn by horses. It is, however, doubtful whether the majority of the horses existing now are derived directly from the indigenous wild horses of Western Europe, it being more probable that they are the descendants of horses imported through Greece and Italy from Asia, derived from a still earlier domestication, followed by gradual improvement through long-continued attention bestowed upon their breeding and training. Such an importation of horses from the East, for the purpose of improving the races of Europe, has taken place at various intervals throughout the whole of the historic period. The most ancient monumental records of Egypt give no sign of the existence of the horse in that country; but about 1900 B.C. (long after the introduction of the ass) it begins to appear, there, as elsewhere, being first employed in drawing chariots used in war and processions.   It was not till a comparatively recent period that the horse was used in agriculture, the ox being almost universally employed in ploughing till the Middle Ages.   The representation in the Bayeux tapestry of a horse drawing a harrow is said to be the earliest indication of the kind, and quite exceptional at that period.

Horses are now diffused, by the agency of man, throughout almost the whole of the inhabited parts of the globe, and the great modifications they have undergone, in consequence of domestication and selective breeding, are well exemplified by comparing such extremes as the Shetland pony, dwarfed by uncongenial climate and scanty food, the thorough-bred race-horse, and the gigantic London dray-horse. The smallest specimens of the former may be not more than half the height of the largest of the latter.*

* Mr. R. Brydon, writing in the *Journal of the Royal Agricultural Society of England*, 3d series, vol. i. part 1 (1890), says: "Having measured many hundreds of them [Shetland ponies], I am convinced that ten hands is the average height, and that very few are found outside a range of from 9.2 to 10.2.   An occasional specimen is met with as low as 8.2 when full grown, but anything under nine hands is extremely rare, and the largest of the pure breed rarely exceed 11 hands." On the other hand, cart-horses between 17 and 18 hands in height are not uncommon.

Perhaps the most striking instance, as it has the certainty of a mathematical demonstration, which can be given of the change of constitution and capability brought about by careful selective breeding in a comparatively short space of time, is seen in the steady progress that has been made in improving the pace of the American fast-trotting horse. Between 1818, when records began to be systematically kept, and 1885, the time for a mile heat has been gradually improved from three minutes (the fastest ever accomplished at the former date, and which previously was not thought possible) to two minutes, eight seconds and three-quarters, which was attained on July 30 of the last-named year. Although this is at present the highest record, past experience renders it probable that it is not the greatest speed ultimately attainable. As bearing upon an important biological problem, much discussed at the present time, it would be extremely interesting to ascertain, if it were possible to do so, whether this result has been acquired solely by breeding from the fastest animals, and so taking advantage of any, even the slightest, variation which occurs in this direction in order to perpetuate the quality in the race; or whether the careful training that the parents have had has been capable of producing a direct influence upon the offspring. The first case would be an illustration of the effects of pure selection ("artificial" in this instance, but coming under the same category of causes of modification as the "natural selection" of Darwin and Wallace), the latter, of inheritance of characters acquired during life, the potency of which has been much called in question of late.

In Australia, as in America, horses imported by the European settlers have escaped into the unreclaimed lands and multiplied to a prodigious extent, roaming in vast herds over the plains where no hoofed animal ever trod before.

The nearest approach to truly wild horses existing at present are the so-called Tarpans, which occur in the steppe-country north of the Sea of Azoff, between the river Dnieper and the Caspian. They are described as being of small size, dun color, with short mane and rounded, obtuse nose. There is no evidence to prove whether they are really wild—that is, descendants of animals which have never been domesticated—or feral—that is, descended from animals which have escaped from captivity, like the horses that roam over the plains of America and Australia, and the wild boars that now inhabit the forests of New Zealand.

Darwin infers that, aboriginally, the horse must have inhabited countries annually covered with snow, for he long retains the instinct of scraping it away with his fore-feet to get at the herbage beneath. Cattle, on the other hand, not having this instinct, perish when left to themselves when the ground is long covered with snow.

Equus przewalskii, Poliakof.—Much interest, not yet thoroughly satisfied, has been excited among zoölogists by the announcement (in 1881) by M. Poliakof of the discovery by the late distinguished Russian explorer, Prejevalsky, of a distinct species of wild horse.* One specimen, unfortunately, only was obtained, while searching for wild camels in the sandy desert of Central Asia near Zaisan. It is described as being so intermediate in character between the equine and the asinine group of *Equidæ* that it completely breaks down the generic distinction which some zoölogists have thought fit to establish between them. It has callosities on all four limbs, as in the horse, but only the lower half of the tail is covered with long hairs, as in the ass. The general color is dun, with a yellowish tinge on the back, becoming lighter towards the flanks and almost white under the belly, and there is no dark dorsal stripe. The mane is dark brown, short, and erect, and there is no forelock. The hair is long and wavy on the

* *Proc. Imp. Russian Geographical Society*, 1881, pp. 1–20, translated by C. Delmar Morgan, in *Ann. Mag. Nat. Hist.* (5) viii. pp. 16–26 (1881).

head, cheeks, and jaws.   The skull and the hoofs are described as being more like those of the horse than the ass.

Until more specimens are obtained it is difficult to form a definite opinion as to the validity of this species, or to resist the suspicion that it may not be an accidental hybrid between the kiang and the horse. *

WILD ASSES.—The remaining existing species of *Equidæ* belong to the asinine group as defined above, and may be conveniently divided into the plain-colored, or true asses, and the striped, or zebras.

The extensive open plains of various parts of Asia, from Syria in the west, through Persia, Afghanistan, the north-west of India, and the highlands of Tartary and Thibet from the shores of the Caspian to the frontiers of China, are the home of numerous herds of wild asses, the individuals in each of which may be from a dozen to twenty in number, or amount to thousands, as described by Dr. Aitchison in his report on the zoölogical results of the Afghan Frontier Expedition of 1884.   They present such a

* The brothers Grijimailo, in a paper published last year in the *Isvestija* of the Russian Geographical Society (of which a translation will shortly appear in the *Proceedings of the Royal Geographical Society of London*), mention meeting with this wild horse in the desert of Dzungaria, and are said to have secured four skins and a skeleton of the species, a full description of which, it may be hoped, will shortly be forthcoming.

general resemblance to each other—being all of a uniform yellowish or isabelline color, lighter or white below, and all having a dark brown stripe along the middle of the back, and usually no cross-stripe on the shoulders—that it is considered by many naturalists that they should all be regarded as belonging to one species—*Equus hemionus* of Pallas. There are, however, such marked differences in size, form, and shade of color, that they may be easily divided into three local varieties, or races, which have been described and named as distinct species. The true *Equus hemionus*, the kiang or dzeggetai, is the largest and the darkest in color, being of a rufous bay, and more approaches the horse in general appearance. It inhabits the high table-lands of Thibet, where it is usually met with at an elevation of 15,000 feet and upwards. Smaller, and paler in color, being sometimes almost silvery-white, is the onager (*E. onager*, Pall.), from Persia, the Punjab, Scinde, and the Desert of Cutch. Differing but slightly, if at all, from this, is the Syrian wild ass, described by Geoffroy under the name of *Equus hemippus*. These three all closely resemble each other in their habits, and are all remarkably swift of foot, having been known to outstrip the fleetest horse in speed. None of them have ever been domesticated.

The origin of the domestic ass (*Equus asinus*,

Linn.), which is nearly as widely diffused and useful to man as the horse, was for long a matter of
uncertainty.   It was known and used in Egypt long
before the horse, and the general belief that it was

Fig. 11.—African Wild Ass (*Equus asinus*) and foal.

*From a photograph by Major J. F. Nott of animals living in the Zoölogical
Society's Gardens.*

first domesticated in that land has been confirmed
by the discovery of a wild ass in Abyssinia and other
parts of the districts of north-eastern Africa lying
between the Nile and the Red Sea which so closely
7

resembles certain breeds of the well-known domestic animal as to leave little doubt as to their identity. This has been called *Equus tæniopus* (band or stripe-footed) by Heuglin, on account of the frequent presence of black, transverse markings upon the lower parts of its limbs.   If its identity with *E. asinus* is admitted, the former name will no longer be required. It differs from the Asiatic species in being of a more pure gray and less rufous or yellowish color, and especially in the presence of a more or less distinct vertical, black mark (sometimes faint and narrow) on the shoulder, corresponding to the stripe so constantly seen in the common domestic animals. Its ears are also of greater length than in the Asiatic species of wild ass.   Sir Samuel Baker says: "Those who have seen donkeys only in their civilized state have no conception of the beauty of the wild or original animal. It is the perfection of activity and courage, and has a high-bred tone in its deportment, a high-actioned step when it trots freely over the rocks and sand, with the speed of a horse when it gallops over the boundless desert."

As with most other animals of the group, its flesh is eaten and much appreciated by the natives of the countries in which it lives. The bray of the Abyssinian wild ass is the same as that so characteristic of the domestic variety, and the marked aversion of the latter to cross the smallest streamlet—an aversion which it shares with the camel—and the evident delight with which it rolls itself in the dust, seem to point to arid deserts as its original home.

THE DOMESTIC ASS is too well known to require description. Although the variations produced by differences of climate, treatment, and breeding are not so great as they are in the horse, they are still considerable, and, if careful selection and improvement had been more attended to, would certainly be far greater. As it is, the continued neglect and ill-treatment to which this unfortunate animal has been too often subjected, as being essentially the servant, or, rather, the slave of the poor man all over the world, has led to deterioration both of its physical qualities and character.

Though gray is the prevailing color of this species, many varieties of that color occur, and instances of every shade between it and pure white on the one hand, and dark brown or black on the other, are met with. The dark, vertical stripe on the shoulder varies much in breadth and intensity of coloring: sometimes it is double, and not infrequently altogether absent. The median dorsal stripe is usually conspicuous. In size, also, there are great differences, the asses used by the lowest caste people of the north of India being scarcely larger than a Newfoundland dog; and in Southern Europe, especially Spain, Italy, and Malta, they are greatly superior; while careful selective breeding in Kentucky has raised their height to 15 or even 16 hands. These large varieties are chiefly in request for the purpose of breeding mules. The milk of the ass, containing more sugar and less caseine than that of the cow, has long been valued as a nutritious diet for persons of weak digestion. Mounteney Jephson says there are great herds of donkeys in a district to the east of the Dinka country, which the natives only use for milking, and not as beasts of burden.*

The ass, unlike the wild horse, is not indigenous in Europe. In England, there is evidence of its presence so early as the reign of the Saxon Ethelred, but it does not appear to have been common till after the time of Queen Elizabeth.

STRIPED MEMBERS OF THE ASININE GROUP OF EQUIDÆ.—These are all inhabitants of the continent of Africa. The animal of this group which was first known to Europeans, and was formerly considered the most common, is the true zebra (*Equus zebra*, Linn.), sometimes called the mountain zebra. It inhabits the mountainous region of Cape Colony, but now, owing to the advances of civilized man into its somewhat restricted range, it has become very scarce,

* *Emin Pasha and the Rebellion at the Equator*, 1890, p. 96.

and is at present limited to a narrow tract near the northern frontier of the colony. A second species, Burchell's zebra (*Equus burchelli*, Gray), still roams in large herds over the plains to the north of the Orange River, but in yearly-diminishing numbers.

FIG. 12.—Common or Mountain Zebra (*Equus zebra*).
From a photograph by Mr. Gambier Bolton of an animal living in the Zoölogical Society's Gardens.

Both species are subject to considerable individual variations in marking, but the following are the principal characters by which they can be distinguished.

Equus zebra is the smaller of the two (about four feet high at the shoulders), and has longer ears, a tail more scantily clothed with hair, and a shorter mane. The general ground-color is white, and the stripes are black; the lower part of the face is bright brown. With the exception of the abdomen and the inside of the thighs, the whole of the surface is covered with stripes, the legs having narrow, transverse bars reaching quite to the hoofs, and the base of the tail being also barred. The outside of the ears have a white tip, and a broad, black mark occupying the greater part of the surface, but are white at the base. Perhaps the most constant and obvious distinction between this species and the next is the arrangement of the stripes on the hinder part of the back, where there are a number of short, transverse bands passing between the median longitudinal, dorsal stripe and the uppermost of the broad stripes which pass obliquely across the haunch from the flanks towards the root of the tail. There is often a median longitudinal stripe under the chest.

Equus burchelli is a rather larger and more robust animal, with smaller ears, a longer mane, and fuller tail. The general ground-color of the body is pale yellowish brown, the limbs nearly white, the stripes dark brown or black. In the typical form the stripes do not extend on to the limbs or tail; but there is great variation in this respect, even in animals of the same herd, some being striped quite down to the hoofs, as shown in the specimen figured

(this form has been named *E. chapmani*).   There is
a strongly-marked median longitudinal ventral black
stripe, to which the lower ends of the transverse side
stripes are usually united; but the dorsal stripe (also

FIG. 13.—Burchell's Zebra (*Equus burchelli*).

*From a photograph by Mr. Gambier Bolton of an animal living in the
Zoölogical Society's Gardens.   The legs are more striped than is usual
in this species.*

strongly marked) is completely isolated in its posterior half, and the uppermost of the broad haunch
stripes runs nearly parallel to it.   A much larger
proportion of the ear is white than in the other
species.   In the middle of the wide intervals of the

broad black stripes of the flanks and haunches fainter stripes are generally to be seen.

This animal is generally spoken of as the "quagga" by colonists and hunters, but it must not be confounded with the species to be described under that name presently. Its flesh is greatly relished by the natives as food, and its hide is very valuable as leather. By far the greater proportion of zebras exhibited in European zoölogical gardens and menageries at the present time belong to this species, and it is frequently bred in confinement, and the attempts made to break it in, and train it for riding and driving, have been attended with partial success.

In 1882 a living zebra was sent from Shoa, a country lying to the south of Abyssinia, to the then President of the French Republic, who deposited it in the Jardin des Plantes, and, being obviously different from any that had hitherto been seen in Europe, it was named by M. Milne-Edwards *Equus grevyi*, in compliment to his political chief. On a white ground-color, it is very finely marked all over with numerous delicate, intensely black stripes, arranged in a pattern quite different from those of the other species. In view of the great variability of the markings of these animals, as long as but one individual of this form was known some doubts were expressed as to whether it might not be an exception-ally-colored individual of one of the other species; but subsequently other specimens, presenting almost exactly the same characters, have been received from Somali-land,* and it seems probable that all the zebras which we know to exist in the northern districts of East Africa belong to this species. The very recent discovery of such a remarkable form of animal, and the imperfect knowledge we possess of its geographical distribution, is a striking illustration of how much still remains to be done before we can consider our information is complete regarding even some of the larger and most conspicuous forms of animal life.

Though zebras have not been found depicted on the Egyptian monuments, they were known to the Romans, and occasionally exhibited in the amphithea-ters, under the name of "*hippotigris.*" Dion Cassius reports that Caracalla exhibited in the circus an ele-phant, a rhinoceros, a lion, and a hippotigris; and as many as twenty are stated to have been collected for the triumph of Gordian the Third, and exhibited by his murderer and successor, Philip the Arabian (A.D. 244).

The QUAGGA, or COUAGGA (*Equus quagga*, Gmelin),

* See Sclater, *Proceedings of the Zoölogical Society of Lon-don*, 1890, p. 413.

is another modification of the zebra group. The color of the head, neck, and upper parts of the body is reddish-brown, irregularly banded and marked with dark brown stripes, stronger on the head and neck, and gradually becoming fainter, until lost on the flanks, the haunches and hind quarters being

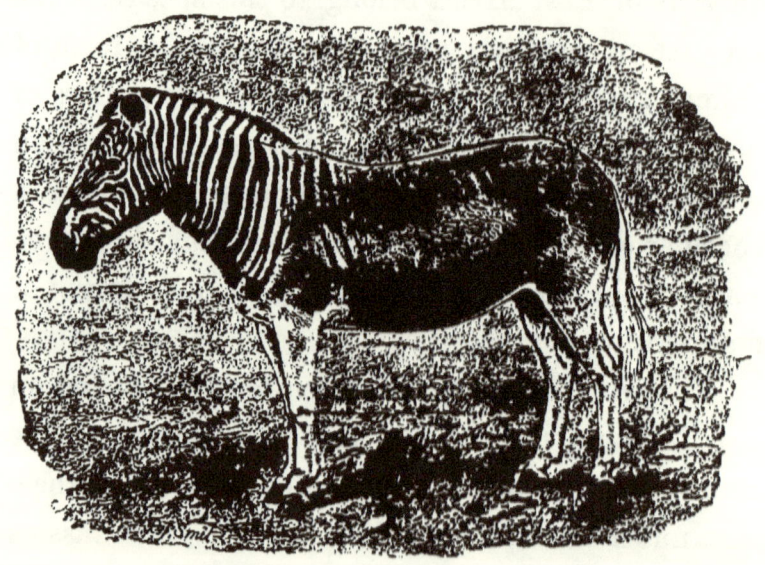

Fig. 14.—Quagga (*Equus quagga*).

From a photograph by Mr. York of an animal which lived in the gardens of the Zoölogical Society of London, 1851-72.

quite free of stripes. There is a broad, dark, median dorsal stripe. The under surface of the body, the legs, and tail, are nearly white, without stripes. The crest is very high, surmounted by a standing mane, banded alternately brown and white. Though never really domesticated, quaggas have occasionally been trained to harness. A pair were driven in Hyde Park, by Mr. Sheriff Parkins, in the early part of the present century. The name is an imitation of the shrill barking neigh of the animal—"ouag-ga, ouag-ga," the last syllable very much prolonged.

There can be little doubt but that, owing to the great improvements in the precision and range of fire-arms, and the general extension of their use into countries where till lately they were unknown, all wild animals which yield any production of value to man, or offer temptations to the sportsman, especially those whose geographical distribution is limited, will soon cease to exist upon the earth. The American bison is one of the most conspicuous instances of rapid extermination of an animal which flourished but very recently in vast numbers, and which, but for the causes just mentioned, might in all probability have continued to exist for long ages. The various species of the large game of Africa are quickly following in the same course. The quagga, although described by Harris in 1839 as existing in "immense herds," is already nearly, if not quite, extinct, the value of its hide being the prime cause of its destruction. Regarding its former geographical distribution, Mr. H. A. Bryden makes the following interesting remarks: "The range of the true quagga was even more arbitrarily

defined. This animal, formerly so abundant upon
the far-spreading karroos of the Cape Colony and
the plains of the Orange Free State, appears never
to have been met with north of the Vaal River. Its
actual habitat may be precisely defined as within
Cape Colony, the Orange Free State, and part of
Griqualand West. I do not find that it ever extended to Namaqualand and the Kalahari Desert to
the west, or beyond the Kei River, the ancient eastern
limit of Cape Colony, to the east. In many countries, and in Southern Africa in particular, nothing
is more singular than the freaks of geographical distribution of animals. A river, or a desert, or a little
belt of sand or timber—none of which, of themselves,
could naturally oppose a complete obstacle to the
animal's range—is yet found limiting thus arbitrarily
the habitat of a species." *

There are thus at least seven modifications of the
horse type, at present or very recently existing, sufficiently distinct to be reckoned as species by all zoölogists, and easily recognized by their external characters. They are, however, all so closely allied that
each will, at least in a state of captivity, cross with
perfect freedom with any of the others. Cases of

---

* *Kloof and Karroo* (*Sport, Legend, and Natural History
in Cape Colony,* 1889).

half-breeds are recorded between the horse and the quagga, the horse and Burchell's zebra, the horse and the hemionus or Asiatic wild ass, the common ass and the zebra, the common ass and Burchell's zebra, the common ass and the hemionus, the hemionus and the zebra, the hemionus and Burchell's zebra. The two species which are, perhaps, the furthest removed in general structure—the horse and the ass —produce, as is well known, mules, which, in some qualities useful to man, excel both their progenitors, and in some countries, and for certain kinds of work, are in greater requisition than either. Although occasional instances have been recorded of female mules breeding with the males of one or other of the pure species, it is doubtful if any case has occurred of their breeding *inter se*, although the opportunities of doing so must have been great, as mules have been reared in immense numbers for several thousands of years. We may therefore consider it settled that the different species of the group are now in that degree of physiological differentiation which still enables them to produce offspring with each other, but does not permit the progeny to continue the race, at all events unless reinforced by the aid of one of the pure forms.

The several members of the group show mental differences quite as striking as those exhibited by their external form, and more than, perhaps, might be expected from the similarity of their cerebral organization. The patience of the ass, the high spirit of the horse, the obstinacy of the mule, have long been proverbial. It is very remarkable that, out of so many species, two only should have shown any aptitude for domestication, and that these, too, should have been from time immemorial the universal and most useful companions and servants of man, while all the others remain in their native freedom to this day. It is, however, still a question whether this really arises from a different mental constitution, causing a natural capacity for entering into relations with man, or whether it may not be owing to their having been brought gradually into this condition by long-continued and persevering efforts, when the need of their services was keenly felt. It is quite possible that one reason why nearly all of the attempts to add new species to the list of our domestic animals in modern times have ended in failure, is that it does not answer to do so in cases in which existing species supply all the principal purposes to which the new ones might be put. It can hardly be expected that zebras and quaggas fresh from their native mountains and plains can be brought into competition as beasts of burden and draught with horses and asses, whose naturally useful qualities have been augmented by the training of thousands of generations of progenitors. It must be remembered also, that the original habitat of both the last-named species probably lay in those countries in which human civilization took its rise, and that they would therefore naturally be the first to be brought beneath its influence.

CHAPTER III.

THE HEAD AND NECK.

NEXT to the body of man, there is none of which
the anatomy has been more thoroughly worked out
and more minutely described than that of the horse.
It is, in fact, the one other animal body that is made
the regular subject of dissection by a whole profes-
sion of students, and to which numerous special
treatises are devoted. Monographs on its structure,
many of them copiously and beautifully illustrated,

abound in most languages of the civilized world.* It might, therefore, seem almost superfluous to add anything further to the subject—certainly difficult to say anything new.

The topographical anatomy of the horse has, however, been always hitherto described just as if it were a complex piece of machinery, isolated and distinct from anything else in the world, the very names given to the parts of which it is composed often having relation only to their conditions of existence in the horse, and being entirely different from those in use for the corresponding parts of man or of other animals. Until lately, at least, the idea that the

* Among the most important of these are:

G. Stubbs: *Anatomy of the Horse*, 1766.

W. Percivall: *The Anatomy of the Horse*, 1832.

E. F. Gurlt: *Anatomische Abbildungen der Haussäugethiere*, 1824; and *Handbuch der vergleich. Anat. der Haussäugethiere*, 1822.

A. G. T. Leisering: *Atlas der Anatomie des Pferdes*, 1861.

Leisering and Müller: *Handbuch der vergleichenden Anatomie der Haussäugethiere*, 6th edit., 1885.

Chauveau and Arloing: *Traité d'anatomie comparée des animeaux domestiques*, 1871; and English edition by G. Fleming, 1873.

M. S. Arloing: *Organisation du pied chez le cheval* (*Ann. Sci. Nat.*, 1867).

Franz Müller: *Lehrbuch der Anatomie des Pferdes*, 1853.

Cuyer and Alix: *Le Cheval*, 1886.

J. McFadyean: *Anatomy of the Horse: a Dissection Guide*, 1884.

W. Youatt: *The Horse*, 1831.

8

peculiarities of the horse's structure are all modifications of a more generalized form, and that their significance can only be understood after a wide study
of the anatomy of allied forms, has never entered
into the mind of any veterinary anatomist.   Certainly, in some of the most recent works, such as
that of Chauveau, attempts to harmonize the nomenclature of parts with that used elsewhere show a
recognition of the community of structure and origin
between the horse and other animals; but still the
knowledge imparted in them has been more adapted
to the technical requirements of the practitioner than
to the enlightenment of those who wish for a broader
and more philosophical view of the ways of nature.

It is only proposed here to select a few of the
most leading parts, which may be of general interest,
and to show their signification and relation, describing them, as far as possible, in language which can
be understood by those who are not professional
anatomists.

For convenience the subject may be divided according to the regions of the body in which the parts
spoken of are placed, certain of those situated in
the head and neck being first selected for consideration; while the limbs, which are of as great importance philosophically as they are practically, will be
reserved for another chapter.

THE SKULL.

The general form of the head of the horse is determined by that of the skull, which forms its supporting framework, and which is of very peculiar and characteristic shape. As in other animals, it is composed of two main portions: (1) the cranium, or skull proper, consisting of a great number of bones, originally quite distinct, but which are eventually firmly united so as to form a solid mass;* and (2) the mandible, or lower jaw, fastened to the former by a freely movable hinge-joint.

The cranium is movably joined to the front end of the vertebral column by means of a pair of oval eminences called "condyles," which fit into corresponding cavities in the atlas, or first vertebra of the neck. Between these condyles is a large opening (*foramen magnum*), through which passes out of the cranium the spinal cord, or backward prolongation of the central nervous system, which is expanded in the head to form the brain.

The cranium may be roughly divided into two portions—a hinder part, or brain-case, consisting of a solid bony capsule for inclosing and protecting

---

* The outlines of the individual bones are perfectly well marked in young subjects, being indicated by fine dividing-lines, called sutures. In old age these often become more or less obliterated by the union of the contiguous bones.

the brain; and a facial part, for the support of the organs of sight, taste, and smell, and of those concerned in seizing and masticating the food.

The skull of a man (Fig. 15) and the skull of a horse (Fig. 16) are composed of exactly the same number of bones, having the same general arrangement and relation to each other. Not only the individual bones, but every ridge and surface for the attachment of muscles, and every hole for the passage of artery or nerve, seen in the one can be traced in the other. Yet they differ remarkably in general aspect. The difference mainly lies in this: in man the brain-case is very large and the face of relatively minute proportions. In the horse, on the other hand, the brain is extremely reduced, and the face, especially the mouth, of enormous size. In other words, the characteristic form of man's head is chiefly due to his great brain, that of the horse to the comparatively large development of the apparatus for masticating his food.

Taking the different regions of the horse's skull (Fig. 16) into closer consideration, and beginning at the hinder, or "occipital" end, we may observe the rounded, almost polished surface of the condyles (oc), already mentioned, which, fitting accurately into the corresponding depressions of the atlas, and in life covered with a soft, perfectly smooth layer of carti-

lage and lubricated with synovial fluid, allow the
head to move freely up and down, or sideways, even

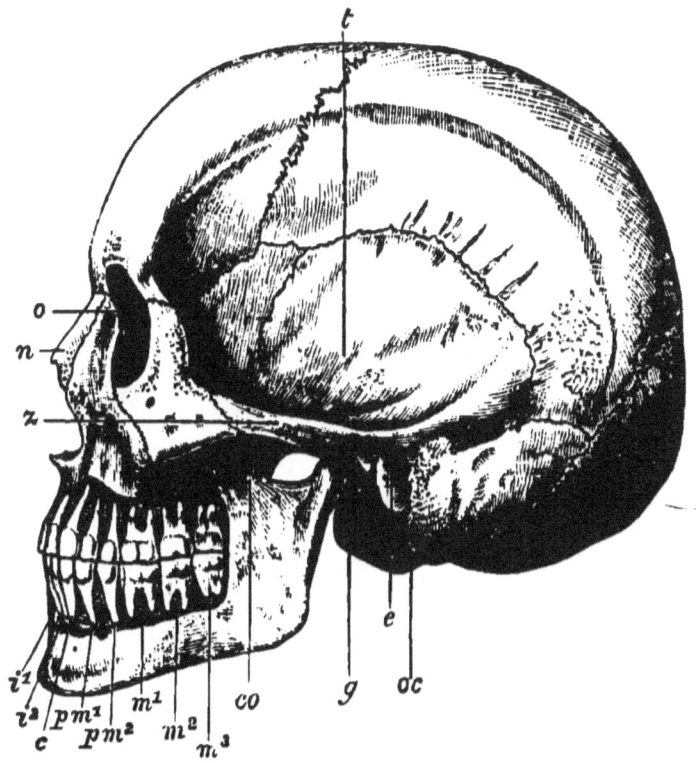

FIG. 15.—Side view of the skull of man, with the
bone removed so as to show the whole of the teeth.
z, zygomatic arch; n, nasal bone; o, or it; t, temporal fossa; oc, occipital condyle; e, external auditory opening; g, glenoid fossa for articulation of
the lower jaw; c, coronoid process of lower jaw;
i^1 and i^2, incisor teeth; cco, canine; pm^1 and pm^2,
premolar teeth; m^1, m^2, and m^3, the three molar
teeth. .

when the neck is fixed. The same region also shows
various roughened, projecting ridges or promontories
for the attachment of the powerful ligaments and

muscles required to support and move so heavy a head, projecting forwards at the end of so long a neck. Above, on each side, are the " occipital crests," joining in the middle line to form the " occipital pro-tuberance," to which that remarkable structure, the "nuchal," or "cervical ligament," to be spoken of further on, is attached. On each side a large, wing-like process (par-occipital) * descends, for the attach-ment of the great lateral muscles of the neck. The head of man, nicely balanced on the top of the verte-bral column, does not require any such great devel-opment of these parts, and they are, consequently, in a quite rudimentary condition in him.

On the lateral surface of the skull, the opening (em) which leads to the internal ear, or true organ of hearing (embedded in the bones which form the side wall of the brain-case), will be seen, to the roughened margin of which the base of the cartilag-inous " pinna," or projecting external ear, is attached. Although we commonly speak of this latter as the " ear," as it is the only externally visible part of the complicated organ by which sounds are recognized, it is a mere accessory, the use of which is to aid in collecting the vibrations passing through the air,

---

* This is the "styloid process" of veterinary anatomy, but not to be confounded with the parts bearing the same name in human anatomy.

and direct them towards the internal, delicate, and beautifully-constructed apparatus in which, by their

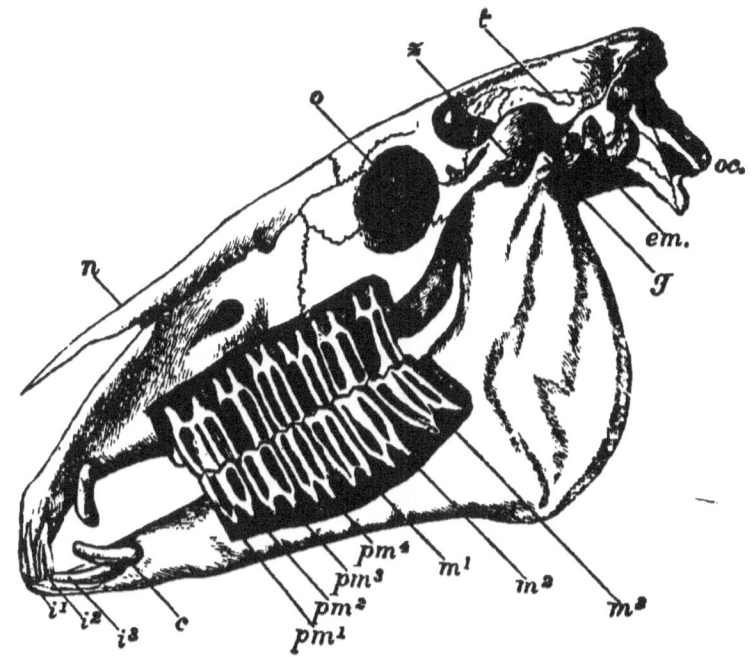

FIG. 16.—Side view of the skull of horse, with the
bone removed so as to show the whole of the
·teeth.   *n*, nasal bone ; *o*, orbit ; *z*, zygomatic arch ;
t, temporal fossa ; *oc*, occipital condyle ; *em*, external auditory opening ; *g*, glenoid fossa for articulation of the lower jaw ; i^1, i^2, i^3, the three incisor
teeth ; *c*, the canine ; pm^1, the situation of the rudimentary first premolar, which has been lost in
the lower, but is present in the upper jaw ; pm^2,
pm^3, and pm^4, the three fully-developed premolar
teeth ; m^1, m^2, and m^3, the three true molar teeth.

effects upon the terminations of the auditory nerve,
they produce the sensation of hearing.

In front of the ear-opening arises a curious bridge
of bone (*z*), which arches forwards to join the skull

again at the hinder part of the face. This is called the "zygomatic arch," and is almost constantly present in the skulls of mammals.

Standing out as it does, it allows the upper part of the under jaw to work beneath it, and its outer surface affords a very advantageous point of attachment to a great muscle (masseter), to be spoken of presently. The side of the brain-case between this and the top ridge of the skull is called the "temporal fossa" (*t*). Though bounded by raised ridges all round, enough to give it the general character of a depression or fossa, its floor is formed by a convex surface, the side of the actual brain-cavity. The fossa is mainly filled in life by one of the muscles (temporal) which close the jaw, but its anterior part contains much fat, the loss of which in old horses gives rise to the characteristic depression seen in them above the eye. Farther forward is the cavity (*o*), of almost circular outline, in which the eyeball is lodged, called the "orbit," with very sharply defined and complete outer and hinder margin, formed by a bridge of bone passing upwards from the zygomatic arch to join the "frontal," or forehead-bone. This is a point in the anatomy of the horse to be especially noticed, as it gives a very characteristic appearance to his skull. The interest of this bridge of bone, dividing the orbit from the temporal fossa, is that it did not exist in any of the primitive Ungulates of the Eocene period, in which, consequently, these two cavities, or fossæ, were freely continuous (see skull of *Phenacodus* in Fig. 2, p. 21). Moreover, it does not exist in any other of the Perissodactyles of the present time (tapir or rhinoceros), but is a special and recently acquired character, developed only in the later stages of the horse group, not being met with in any of the ancestral forms until after the close of the Miocene period.

The horse, however, is by no means alone among mammals in possessing it; but whenever it occurs, it appears to be an evidence of advance in structure, being found in the higher and more recent forms of several groups, the lower and more ancient members of which do not possess it. In man it has attained its most complete development, for not only is there a bridge, but also a wall extending inwards from it, cutting off almost completely the two cavities from one another.

In front of the orbit a great, flat expanse, the " cheek," extends quite to the fore-part of the face, giving room for the long row of upper molar teeth, and within, for the lodgment of the highly-developed organ of smell. This is roofed over above by "nasal bones" (*n*) of great size, terminating in front by freely-projecting, pointed, and somewhat decurved ends, which support the well-developed external nostrils. The palate, which forms the floor of all this part of the skull, is remarkable for its great length and comparative narrowness. The front end of the upper jaw consists of the united "premaxillary bones," which expand and curve down to form the semicircular border supporting the large incisor teeth. In the middle line, in front, between the premaxillæ below and the nasals above, is the large, irregular opening of the "anterior nares," leading into a great chamber or passage, divided into two by a vertical median wall or septum. Through this chamber the air passes in respiration to the "posterior nares," a smaller opening at the base of the skull behind the palate, and in the upper part of it is placed the sponge-like mass of bones which support the terminations of the olfactory nerves, constituting the organ of smell. Before leaving the cranium, the " glenoid " cavities, or, rather, surfaces (g), to which the two branches of the mandible are articulated, must be mentioned. They are placed just below the hinder end of the zygomatic arch. They are wide transversely, concave from side to side, convex from before backwards in front and hollow behind, and bounded posteriorly at the inner part by a prominent "post-glenoid" process, which effectually prevents the jaw from being dislocated backwards.

The lower jaw, or "mandible," consists of two halves or branches ("rami"), originally distinct, but firmly united in adult horses by their front ends (the symphysis). Each is articulated to the corresponding glenoid surface of the cranium by its "condyle," placed at the hinder and upper end of the ramus. The smooth, articular surface of this is very wide transversely, but narrow and convex from before backwards. The principal action at this joint is that of a simple hinge, but the form of the contiguous surfaces allows a certain amount of motion in other directions, far more, for instance, than is permitted in the very complete interlocking hinge-joint of the Carnivora.

In front of the condyle, and separated from it by a notch, rises a somewhat small and slender, backward-curving "coronoid process" (co, Fig. 15), for the attachment of the temporal muscle, which aids in closing the jaw. Below it is a flat, broad, expanded surface, reaching down to the "angle" (where the horizontal and vertical or ascending portions of the jawbone meet), for the attachment of the huge masseter muscle, arising from the zygomatic arch, and from a well-marked ridge running horizontally forwards on the cheek in continuation of the lower border of the arch. This muscle is the main agent in closing the mouth, and therefore in crushing the food between the molar teeth. The horizontal portion of the jaw, long, straight, and flattened from side to side, carries the great molar teeth, and gradually narrows towards the symphysis, where it expands laterally, to form, with the united opposite ramus, the wide, semicircular, shallow alveolar border for the lower incisor teeth.

The Teeth.

The next parts to which attention may be called are the teeth, which in the horse, though founded upon the same general type as the primitive Ungulates of the Eocene period, have undergone a remarkable amount of specialization, fitting them in an eminent degree for the purpose they have to fulfill.

Number of the Teeth.—For convenience of description teeth are divided, according to their situation in the mouth and other characters, into four sets, called (beginning from the front) *incisors, canines, premolars*, and *molars*. As mentioned in the first chapter, all the early Ungulate mammals, without exception, had on each side, above and below, three incisors, one canine, four premolars, and three molars—that is, eleven on each side above, and eleven below, or forty-four altogether. The modern horse has nearly, if not quite, this full number. The front teeth, or incisors, are the same—six above and six below, taking the two sides together (Fig. 16, i^1, i^2, and i^3). The canines, or "tushes" (c), are present, as a rule, only in the males. The cheek-teeth $(pm^2$ to $m^3)$, or premolars and molars taken together (for there is very little to distinguish them in form or size), are generally but six, instead of seven, on each side above and below. Here, then, is a case of specialization by suppression. One of the teeth of the ancient forms has disappeared. Which is it? The examination of a series of fossil remains shows us that the first of the series—the anterior premolar (pm^1), a fairly large and well-developed tooth in Phenacodus and Hyracotherium—gradually became smaller and smaller as time advanced. It is still present in Anchitherium, sometimes present and sometimes absent in Hipparion. But has it entirely disappeared in the modern horse? What do we read in old books on veterinary surgery? "Wolves' teeth are two very small, supplementary teeth, appearing in front of the molar teeth, and supposed to have an injurious effect on the eyes (!), and are, therefore, often removed by farriers."

These little rudiments of teeth, about which such nonsense as the above has been written, are, when properly understood, of intense interest. Their diminutive size, their irregular form and inconstant presence, combined with their history in the extinct horse-like animals, show them to be teeth which, for some reason to us at present unknown, have become superfluous—have been very gradually and slowly (as in the case of all operations of the kind) dispensed with, and are in the stage to which the horse has now arrived in its evolution, upon the point of disappearance. The presence of these so-called "wolves' teeth" alone is sufficient, if we had no other proof, to show that the horse is not an isolated creation, but one link in a great chain of organic beings. The fact that these teeth are almost always met with in the upper jaw only, should be noted in connection with what has been previously mentioned respecting the dentition of the tapir. The first upper premolar is retained in that animal, while the corresponding lower tooth has entirely disappeared.

It would be very interesting, if a sufficiently large number of specimens could be examined, to obtain some statistical imformation as to the relative frequency of the occurrence of these teeth in the different species of wild and different breeds of domestic horses. They are usually so loosely attached in the skull that they become lost in specimens prepared for museums; but indications can generally be seen on the bone, if they have been present.

General Characters and Structure of Teeth.—Before describing the teeth of the horse a little more in detail, it will be necessary to give some slight account of the characters and structure of these organs in general, in order that the special descriptions may be better understood.

Every tooth may be divided into two principal parts, a "crown," and a "root" (sometimes erroneously called "fang"). The part connecting the two, often indicated by a constriction, is called the "neck." The crown is the only part which is seen in the living animal, the root being implanted in a socket in the bone, just as the roots of a tree are in the ground. The crown may be variously shaped—conical and pointed, chisel- or awl-shaped, broad, flat, or rounded; or it may be complicated by the development upon its surface of elevations or tubercles, called cusps, or by variously disposed crests or ridges. The root may be single, or divided into two or more conical, tapering branches.

In structure, the teeth are composed of several distinct substances, differing from each other in character and degree of hardness. The most important of these are:

1. The *pulp*, a soft substance, abundantly supplied with blood-vessels and nerves, constitutes the central axis of the tooth, and affords the means by which its vitality is preserved. This occupies a larger relative space, and performs a more important purpose in the

young, growing tooth than afterwards, as, by the calcification and conversion of its outer layers, the principal hard constituent of the tooth, the dentine, is formed.   In teeth which have ceased to grow the pulp occupies a comparatively small space, which in the dried tooth is called the pulp-cavity.   This communicates with the external surface of the tooth by a small aperture at the apex of the root, through which the branches of the nutrient blood-vessels and sensitive nerves necessary to maintain the vitality of the tooth pass in, to be distributed to the pulp.   In growing teeth the pulp-cavity is widely open below, while in advanced age it often becomes obliterated, and the pulp itself entirely converted into bone-like material.

2. The *dentine*, or *ivory*, forms the principal constituent of the greater number of teeth.   It is a very hard but elastic substance, white, with a yellowish tinge, and slightly translucent.   Its chemical composition is very like that of bone, but its microscopical structure is altogether different.

3. The *enamel* constitutes a thin investing layer, complete or partial, of the exposed or working surface of the dentine of the crown of the teeth of most mammals.   This is the hardest tissue met with in the body, containing from 95 to 97 per cent. of mineral substances (chiefly phosphate, and some carbonate of

lime, with traces of fluoride of calcium). Enamel is easily distinguished from dentine with the naked eye by its clear, bluish-white, translucent appearance.

4. The *cement*, or *Crusta petrosa*, is always the most externally placed of the tissues of which teeth are composed. It is often only found as a thin layer upon the surface of the root; but sometimes, as in the complex-crowned molar teeth of the horse and elephant, it is a structure which plays a very important part, covering and filling in the interstices between the ridges of the enamel. Its structure and chemical composition is almost exactly that of ordinary bone.

Succession of Teeth.—The dentition of all mammals consists of a definite set of teeth, of constant and determinate number, form, and situation, and, with few exceptions, persisting in a functional condition throughout the natural term of the animal's life. In many species these are the only teeth which the animal ever possesses—the set which is first formed being permanent, or, if accidentally lost, or decaying in extreme old age, not being replaced by others. But in the horse, as in all other Ungulates, as well as in man, and, in fact, the majority of the class, certain of the teeth are preceded by others, of a smaller size, which occupy the place of the permanent teeth during the growth and gradual

9

maturation of the latter, and especially while the jaws are acquiring size and strength sufficient to support them. In all cases these teeth disappear (by the absorption of their roots and shedding of the crowns) before the frame of the animal has acquired complete maturity. As the first set of teeth are, as a general rule, present during the period in which the animal is nourished by the milk of the mother, the name of "milk-teeth" (French, *dents de lait;* German, *Milchzähne*) has been commonly accorded to them, although it must be understood that the time of their duration has nothing to do with that of lactation. "Temporary teeth," or "deciduous" teeth, are, perhaps, therefore, better names. No mammal has more than two sets of teeth.

Special Characters of the Teeth of the Horse.—Incisors.—To return to the teeth of the horse. The incisors, or "nippers," as they are called in veterinary language, of each jaw, are placed in close contact, forming a semicircle. The crowns are very large, somewhat chisel-shaped, and of nearly equal size. They have all a peculiarity not found in any other existing mammal,* and seen only in the *Equidæ* of comparatively recent geological formations. In the most primitive species these teeth

---

* *Macrauchenia,* an extinct South American Perissodactyle, had somewhat similar incisors.

were simple, and chisel- or awl-shaped. When their crowns became worn in consequence of long-continued use, they presented an external ring of enamel, surrounding a core of the dentine, or ivory, of which the bulk of the tooth is composed. This is the condition of the incisor teeth in the great majority of mammals. The first modification from this simple form consisted in the development of a ridge along the hinder border of the base of the crown, as seen in Fig. 17. There was then a groove between this ridge and the rest of the tooth. By the continuous

FIG. 17.—Incisor tooth of *Anchitherium aurelianense*.

FIG. 18.—Unworn crown of temporary incisor tooth of young horse.

increase of the ridge, and its union with the edges of the main part of the crown on each side, the groove became converted into a deep pit (*infundibulum*), the orifice of which is transversely elongated, and placed behind and rather below the cutting-edge of the tooth. This is the condition seen in a colt's incisor which has just cut the gum (Fig. 18). As wear takes place, the surface, besides the external enamel layer, as in the ordinary simple tooth, shows, in addition (see Fig. 19), a second, inner ring of the same substance surrounding the pit, which, of course, adds greatly to the efficiency of the tooth as an organ for biting tough, fibrous substances. The bottom and sides of the pit are partially lined with cement, but a considerable cavity remains, generally filled, in the living animal, with particles of food, and, being conspicuous from its dark color, it constitutes the "mark" by which the age of the horse is judged. In consequence of its only extending to a certain depth in the crown, it becomes obliterated as

Fig. 19. — Incisor tooth of horse, with the crown partially worn, showing the pit surrounded by its enamel layer, outside of which is the dentine with its external enamel covering. The enamel is represented white, the dentine gray, and the pit black.

the tooth wears away, the section of which then assumes the character of that of an ordinary incisor, consisting of only a core of dentine, surrounded by an external enamel layer. The flattened, worn surface of an incisor tooth, as seen in Fig. 19, is called, in works of descriptive veterinary anatomy, "the table." It is totally different in appearance from the summit of an unworn tooth, with its thin, rounded, shell-like margins surrounding the deep, open cavity, as in Fig. 18.*

The permanent incisors of the horse (like the molars, as will be explained hereafter) differ from those of most mammals in the great length of their crowns, which do not remain fixed in position when they have once come into place, but continue to push up from below, as they wear away at the exposed surface, for a considerable part of the life of the animal. The upper part of the tooth, or that which first appears, is very wide transversely, and narrow from before backwards; but the form gradually alters, becoming narrower from side to side, and finally somewhat triangular in section, flat in front, and with a projecting ridge in the middle, behind. Consequently, the shape of the table alters as the wear of the tooth proceeds, and by its form gives indications of great assistance in determining the age of the animal. A considerably worn table shows, in addition to the mark caused by the pit or

* This appearance has given rise to the term of "shell" teeth, applied to newly-cut, unworn incisors.

infundibulum, another spot, having a cloudy-yellow color, always situated in front of the pit when the two coexist, and continuing, after the obliteration of the former, quite to the base of the root. This is caused by the pulp-cavity, which has become filled up by an irregular deposit of dentine. As it has no surrounding of enamel, it cannot be mistaken for the pit, or true mark.

The three incisors of each side of the jaw, beginning at the middle line, are spoken of as "central" or "pincers," "lateral" or "intermediate," and "corner" teeth. For brevity of description they are symbolized as i^1, i^2, and i^3.

The characters of the incisor teeth in the three existing families of Perissodactyles offer an interesting subject for consideration. All originating in a similar, and comparatively simple form, they have all varied from it in totally different directions. Those of the tapirs show the least change from the primitive condition; those of the rhinoceroses have dwindled down in number and size, to complete disappearance in some species; those of the horses have undergone changes leading, finally, to a complication of structure unknown in any other existing animal. There can be little doubt but that these changes have all been in adaptation to some peculiarities of the environment of the animals, and that each has been best adapted for the purpose which it has had to fulfill; but the relations between use and structure are often of such a delicate and intricate character that they quite escape the recognition of our limited powers of observation.

The Canines.—Separated from the incisors by a short interval are the teeth called in the general language of zoölogy "canines," but usually spoken of in the horse as the "tusks," or, more often, "tushes" (Fig. 16, *c*). They correspond exactly with the tusks of the boar and the great corner teeth of the lion and dog; but in all the *Equidæ* they play a very subordinate part, not being required either as a means of defense or for the purpose of seizing prey. Following a very general rule among the Mammalia, especially marked in the Ungulates and other groups (as monkeys) in which these teeth are not a necessity for procuring food, they are much more developed in the male than in the female. Indeed, they are practically absent in the latter sex, as, when they do occur as an exception, they are in a more or less rudimentary condition. As the canines were present in both sexes in the Eocene and Miocene Ungulates, their loss in the females of the existing *Equidæ* must be reckoned, like the loss of the anterior premolar, among the numerous instances of specialization which this group has acquired.

It may be remarked in passing, that the canines are the only teeth which afford indications by which the sex of an animal may be distinguished, except, of course, such as may be inferred from the general disparity of size which characterizes the entire dentition, in common with the rest of the organization, in many cases.

In the adult male horse they are always present in both upper and lower jaw, but they are smaller than the incisors, and of different form, being, when unworn, pointed at the apex, and presenting nothing resembling the pit or infundibulum. They have a tendency to curve, the concavity being turned backwards. Their outer surface is smoothly convex; their inner surface has a prominent, rounded, longitudinal ridge, and a groove on each side. The borders separating their outer and inner surfaces are, when unworn, sharp and cutting, and meet at the apex.

Diastema.—Isolated as the canines are from the incisors in front, they are separated by a still wider interval (or "diastema") from the molar teeth behind. This toothless interval, called the "bar" in the lower jaw, is of essential importance in the domesticated horse to his master, as without it there would be no room for the insertion of the special instrument of subjugation to his commands—the bit. In the most primitive condition of dentition there appears to have been no such interval, all the teeth being in contact; and this condition is retained, or perhaps regained, by man, almost alone among existing mammals. Already, in Phenacodus, there was an indication of this diastema, and throughout the whole series of Perissodactyles which lead up to the *Equidæ* there has been a gradual increase of its length.

Molar Teeth.—The cheek-teeth, or molars, excluding the rudimentary and inconstant anterior premolar, spoken of before, are six in number, above and below, on each side (see Fig. 16, pm^2 to $\underline{m^3}$). They are all in such close contact, by broad surfaces fitting tightly against each other, that they form together one solid mass, presenting a grinding-surface composed of substances of various densities, and therefore projecting at slightly different levels, interwoven in such an intricate pattern as to form one of the most efficient natural millstones imaginable.

A distinction must be pointed out among these teeth. In a great many animals their form differs so much that they are readily separated into an anterior set of simple character, and a posterior set, larger, broader, and with more complex crowns and roots; and when it was discovered that these also presented a constant difference in their mode of development—the first set being preceded in their places by other teeth of the milk, or deciduous series, and the last set coming up behind the last of the milk-teeth, without any predecessors—the distinction was thought to be of sufficient importance to give them different names, the first being called "premolars," or "false molars," and the last "true molars."

In the horse there is no difference in form or size between the premolars and true molars, and it is only by the analogy of other animals, and by a knowledge of their early history in the horse itself, that we can divide them, and know that the great mill-like mass of cheek-teeth consists of three premolars and three molars.

It is characteristic of a primitive condition of dentition that premolars and molars should differ in form. Such a condition is, doubtless, best adapted for an omnivorous or generalized mode of feeding; but it is a specialty on the part of the Perissodactyles, which was acquired very early in their history, and is, no doubt, in accordance with their strictly vegetarian life, that the premolars have taken on the form of the true molars, and have become as completely adapted as the latter for the grinding function.

Another and still more important deviation which has taken place in the general condition of the molar teeth from their primitive state is this. The crowns of all teeth in the early forms were very low, or short from above downwards, and therefore but slightly elevated above the surface of the jaw. There was a distinct constriction—the neck—between the crown and the root, and when the tooth was developing, as soon as the neck once rose fairly above the margin of the bone, the tooth remained permanently in this position. The term "brachydont" expresses this condition of tooth, which was universal in Perissodactyles up to and including the Anchitherium of the Miocene epoch (see a, Fig. 20). The free surface of the crown presented cusps and ridges upon it, with valleys between; but the valleys were shallow, and had no deposit of cement filling them, the whole exposed surface of the unworn tooth being formed of enamel. When the ridges became worn down by the friction of hard particles of food interposed between the opposing teeth, the dentine of the interior was exposed, forming islands surrounded by enamel. As the wear continued and reached the bottom of the valleys, all the enamel coating disappeared from the upper surface, and nothing remained but a plane surface of comparatively soft dentine, surrounded at the circumference by the enamel. With the progress of time, however, individual succeeded individual, in each of which, probably by insensible degrees, the crowns of the teeth became longer, the valleys deeper, and the ridges not only more elevated, but more curved and complex in arrangement. To give support to these high ridges, and to save them from breaking in use, the valleys and cavities between them became filled up to the top with cement, which was also packed round the whole outer surface of the enameled crown, and as the tooth wore down the result was an admirable grinding-surface, consisting of patches and islands of the two softer substances —dentine and cement—separated by variously re-duplicated and contorted lines of intensely hard en-amel, the greater resistance of which to the attrition of the food caused it to project slightly above the rest of the surface (see section of the tooth of Hip-parion, Fig. 10, p. 71). To this lengthened form of crown the term "hypsodont" is applied. Instead of contracting into a neck and forming roots, its sides continue parallel for a considerable depth in the socket, and as the surface wears away the whole tooth slowly pushes up, and maintains the grinding-edge constantly at the same level above the alveolar bor-der, much as in the perpetually-growing front teeth of many rodents, which never contract at the base to form roots, but continue throughout the life of the animal to grow from below to the same extent as they are worn away at the outer, or cutting-edge. But the horses have not quite attained this condition.

There is still a limit to the growth of their teeth. After a length is attained which, under normal conditions, supplies sufficient grinding-surface to last the lifetime of the animal, a neck and roots are formed, and the tooth is reduced to the condition

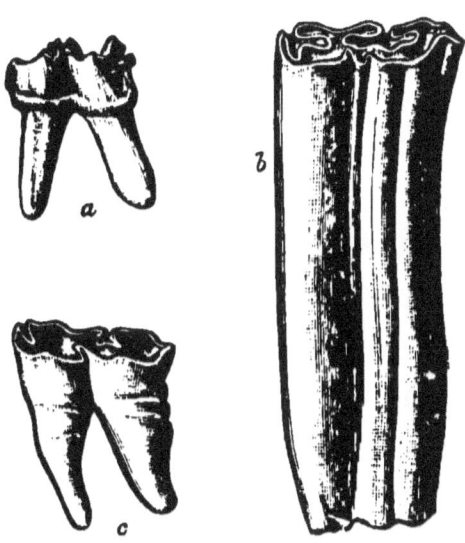

FIG. 20.—a, Lower molar of Anchitherium (brachydont form); b, lower molar of a young horse, with the crown slightly worn and the roots not yet formed (hypsodont form); c, the same tooth of an old horse, with the crown almost entirely worn away and the roots fully formed.

of that of the brachydont ancestor (see b and c, Fig. 20).

It is perfectly clear that this lengthening of the crown adds greatly to the power of the teeth as organs of mastication, and enables the animals in which it has taken place to find their sustenance among the comparatively dry and harsh herbage of the plains, the stalks of which often contain much hard mineral matter, instead of being limited to the soft and succulent vegetable productions of the marshes and forests in which the primitive brachydont forms of Ungulates mostly dwelt.

The hypsodont, or high-crowned type of tooth, which may be looked upon as an intermediate condition between the rooted and the ever-growing type, is by no means peculiar to the molars of the horse. It occurs, as already mentioned, in the incisors of the same animal. It is also met with, in various degrees, in the more recently-developed forms of the rhinoceros family (though not in the tapirs), and in some of the most specialized of the Artiodactyles, as the ox and the sheep, though not attaining in those animals to the same development which it does in the horse.

As there are some differences in the details of the structure of the premolars and molars of the upper and lower jaws, it will be necessary to describe them separately.

Of the six principal teeth which constitute the upper molar series, the four middle ones (the last two premolars, pm^3 and pm^4, Fig. 16; and the first two molars, m^1 and m^2) are almost exactly alike in size, form, and structure, being, roughly speaking, four-sided prisms with a nearly square section. The foremost, pm^2, and the hindermost, m^3, differ from the others, being more triangular in section, the apex of the triangle pointing forwards in the first and backwards in the last.

To understand properly the arrangement of the enamel folds and of the dentine within, and the

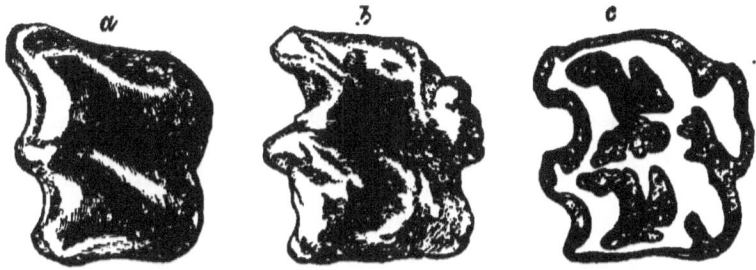

FIG. 21.—*a*, grinding-surface of unworn molar tooth of Anchitherium; *b*, corresponding surface of un-worn molar of young horse; *c*, the same tooth after it has been some time in use. In the latter, the uncolored portions are the dentine or ivory, the shaded parts the cement filling the cavities and sur-rounding the exterior. The black line separating these two structures is the enamel, or hardest con-stituent of the tooth.

cement on the outside of them, as seen in a section or in the naturally worn surface of one of the mid-dle teeth, it is necessary to examine it in its unworn and even unfinished state, before the thick coat of cement has been deposited around the ridges and projections of the surface. Such teeth can always be found within the sockets of the prepared skulls of young animals, and deserve careful study, on account of the light they throw upon the structure of the organ in its maturity. It will be seen (Fig. 21) that the pattern presented by the free surface is essentially similar to that of the corresponding tooth of Anchitherium, which is itself a modification of that of Hyracotherium, from which it is but an easy transition to Phenacodus, as shown in the first chapter. The main difference is that, in the short-crowned tooth of Anchitherium, the ridges and valleys are necessarily very shallow, with sloping walls, and there is no need for a packing of cement around and within them ; while in the horse they descend the whole depth of the elongated crown of the tooth, with nearly parallel walls, so that any part presents an almost identical section, and they are filled in and packed round with an abundance of cement. The four original main cusps—antero-internal, postero-internal, antero-external, and postero-external — and also the two intermediate cusps are distinctly recognizable, but they are prolonged anteriorly and posteriorly into ridges or walls taking a generally crescentic form, with the concavity of the crescent looking outwards, and its convexity inwards. It is this disposition which gives the outer surface of the teeth its deeply ridged and grooved or fluted character, the two deep grooves corresponding with the concavities of the two outer crescents. The internal columns, especially the anterior one, have a great tendency to detach themselves from their crescents formed by the intermediate cusps. The amount of detachment and the form of this column in section are important in determining the various species of fossil *Equidæ*, and its complete detachment in the Hipparion (*Hippotherium*) forms one of the principal characters (already alluded to) by which that genus is distinguished from the other members of the group (see Fig. 10, p. 71).

The two deep holes, of a roughly crescentic shape, filled with cement, are spoken of, for descriptive purposes, as the anterior and posterior lakes. The sinuosities of their enamel margins, which are sometimes extremely complex, present great variations in different species, as also do the indentations in the edges of the sinus which runs forwards from the inner side of the tooth between the two internal columns, the form of the folds at the bottom of which constitutes the only easily recognizable distinction between the molar tooth of the common horse and the ass.

Though the length of the combined grinding-surfaces of the upper and lower molar teeth is practically the same, the latter are scarcely more than half the width from side to side, and present quite a different pattern. As in the upper series, those placed

10

at either extremity narrow off anteriorly and posteriorly, but the four middle ones are almost identical. The pattern of these teeth resembles essentially that of most of the other Perissodactyles. They
consist mainly of two crescents, one placed in front
of the other, with their convexities outwards; but it
is peculiar to the *Equidæ* to have the inner ends of
the crescents complicated by the addition of columns
or lobes, which add considerably to the intricacy of
the pattern exposed when the tooth is worn. The
extremely hypsodont condition and the quantity
of cement which everywhere overlays the enamel
and fills in the interstices of the foldings are also
special characters of this group, which they share
only with the *Elasmotherium* among the Rhinocerotidæ.

The surfaces of the upper and lower molars in
wearing against each other do not come in contact
in a plane horizontal to the long axis of the tooth,
but in one slanting from without upwards, the wear
being greatest on the inner side of the upper teeth
and the outer side of the lower teeth.

The roots of the molars are short, and in the upper ones four in number, except in the first and last,
which have only three. In the lower teeth there are
two, one anterior and one posterior, in position.
After they are fully formed, the tooth does not con-

tinue to grow, but gradually rises towards the surface of the socket, the bottom of which fills up with bone, so that in very old horses the crowns are entirely worn away, and nothing but the roots of the teeth remain, loosely implanted in the jaw. If from any accidental cause one of the teeth is absent, the tooth opposite to it in the other jaw, having nothing to wear against, will gradually rise high above the level of its fellows.

Temporary or Milk-Teeth.—The first, temporary, deciduous, or milk set of teeth of the horse, though on the whole resembling the permanent set, having the characteristic enamel foldings arranged on the same general principle, present several interesting peculiarities.

The incisors are in number the same as the permanent teeth of this class. They are, however, not only smaller, but, as they are only required to be in use for a limited time, they have not need of the lengthened crowns passing indefinitely into the root possessed by the latter, and therefore show distinctly the broad crown, contracted neck, and definite root of the brachydont type of tooth. In this respect they resemble the ancestral form from which the permanent teeth have been derived. The infundibulum or pit is present, but of comparatively little depth.

The canine teeth of the horse evidently belong to the permanent set, not coming into place until the animal is full grown, and lasting throughout its lifetime. It is commonly stated that they have no deciduous predecessors. On this subject, however, the following observations of Lecoq are important:*

"The canine teeth are not shed, and grow but once. Some veterinarians, and among them Forthomme and Rigot, have witnessed instances in which they were replaced; but the very rare exceptions cannot make us look upon these teeth as liable to be renewed. We must not, however, confound with these excep·tional cases the shedding of a small spiculum, or point, which, in the majority of horses, precedes the eruption of the real tusks."

These spicules are in all probability the true milk canines in an extremely vestigial condition; their loss, in the gradual process of degeneration of these teeth, taking place, as might be expected, before that of their permanent successors. This subject would well repay a fuller investigation than it has hitherto met with, as it seems to be another of the numerous instances of rudimentary structures in the

* Quoted in Fleming's translation of Chauveau and Arloing's *Comparative Anatomy of the Domesticated Animals* (1873), p. 352.

horse, pointing to a different condition in the ancestral state.

The diminutive first premolars should probably be regarded as teeth of the permanent set, and, considering how near they are to disappearance, they could hardly be expected to have milk predecessors, especially as such are frequently absent in other animals in which these teeth are fairly well developed.

The functional milk molars are three in number, corresponding in succession with the three functional premolars of the permanent set. The middle one resembles the intermediate permanent molars, but the first and third have their extremities somewhat narrowed, so that the grinding-surface of the whole block presents a representation on a smaller scale of that of the permanent set. The crowns are comparatively short, and distinct roots are formed by the time the growth of the tooth is complete. As the permanent teeth rise up below them these roots are absorbed, and nothing remains but the worn base of the crown, which is finally cast off as its successor becomes fitted to take its place.

Time of Appearance and Order of Succession of the Teeth.—The eruption or cutting through the gums of the temporary teeth commences at about the time of birth, and is complete before the end of the first year, when the young animal has its full set of

twenty-four teeth, three incisors and three molars
above and below on each side of each jaw. The
upper teeth, as a rule, appear somewhat earlier than
those of the lower jaw. Within a very few days
after birth the central incisors make their appearance,
and by the end of the second week they are fairly
up in the mouth. The first and second molars come
into place about the same time. Between the first
and second month the second (lateral) incisors appear, then the third molar, and finally (at about
nine months) the third (corner) incisors, which complete the milk dentition. Of the permanent teeth,
the first true molar appears about the end of the first
year, followed by the second molar before the end of
the second year. These teeth are thus in place before any of the milk-teeth have been shed. At about
two and a half years the first premolar replaces its
predecessor. Between two and a half and three
years the first permanent incisor appears. At three
years the second and third premolars and the third
true molar have appeared, at from three and a half
to four years the second incisor, at four to four and
a half years the canine, and finally, at five years,
the third (corner) incisor, completing the permanent
dentition. Up to this period the age of the horse is
clearly shown by the condition of its dentition, and
for some years longer indications can be obtained

from the wear of the incisor teeth,* though this depends to a certain extent upon the hardness of the
food and other accidental circumstances.

THE LIPS.

The lips of the horse are remarkably sensitive and
flexible. They can be stretched out in various directions, and are much used in gathering food into the
mouth. Any one who has seen a horse take a small
piece of sugar from a child's hand will appreciate the
delicacy and efficiency of these organs as instruments
of prehension. They present a great contrast to the
thick rigid lips of the ox, in which animal the tongue
plays a more important part in the duty of obtaining
food. Flexible and prehensile lips are characteristic
of the Perissodactyles. In most species of rhinoceros
the upper lip is prolonged to a point in the middle
line, which acts almost like a finger, and in the tapirs
it joins with the nose to form a flexible and very mobile snout or short proboscis.

THE NOSTRILS.

The nostrils of the horse are large and very dilat-
able, allowing of the admission of a greater or less

* These are very fully described and illustrated in a
pamphlet called *Dentition as indicative of the Age of the Animals of the Farm*, by Professor G. T. Brown, 2d edit. 1889.

amount of air, according to the demands of respiration. Owing to the great length of the soft palate and its relation to the upper end of the windpipe, breathing takes place entirely through the nose. When men, dogs, and many other animals, in consequence of any great exertion, begin to pant, and require an additional quantity of air to that which is ordinarily taken in by the nose, the mouth comes to the aid of that channel, and is widely opened; but the horse under the same circumstances can only expand the margins of the nostrils, for which action there is a very efficient set of muscles, acting on the cartilaginous framework which supports them and determines their peculiar outline. The variations in the form and amount of dilatation of the nostrils give great character and expression to this part of the horse's face.

Immediately within the margin of the upper part of the nostril is a structure of very considerable interest, which is generally supposed to be peculiar to the horse and its immediate allies, the use of which is entirely unknown. It is a blind pouch, three to four inches in depth, conical in form, though slightly curved, and lying in the cleft seen in the dried skull between the nasal and premaxillary bones. It is a diverticulum from the nasal passage, with which it freely communicates below, and is lined by a continuation of the same smooth mucous membrane which lines the passage. In veterinary anatomy it is called the "false nostril."

If this were all we knew about this organ it would be unsatisfactory enough, but it immediately acquires interest when we learn that in the tapir a similar structure, only in a very much more developed condition, is found. In that animal it runs upwards, as a long, narrow tube, from the external nostril, at first in contact with its fellow of the opposite side, and afterwards, taking a curiously curved course, terminates in a dilated, closed extremity, which lies in a distinct groove by the side of the upper part of the nasal bone. Its walls are cartilaginous, and convoluted in such a manner as greatly to increase the area of the internal surface. It is obvious that the "false nostril" of the horse cannot be looked upon as anything specially belonging to the economy of that animal, but rather as a rudimentary condition or survival of a structure which is far more highly developed in some of the more primitive forms of Perissodactyles. This view is greatly strengthened by the recent discovery of an exactly similar structure in the rhinoceros, only in a condition intermediate between that in which it is found in the horse and the tapir.*

* E. F. Beddard, *Proceedings of the Zoölogical Society of London*, 1889, p. 10.

Thus, an organ which, when only known in one animal, appeared strange, anomalous, and puzzling, because there seemed nothing to account for its presence, acquires in the light of wider knowledge a much deeper interest; for if we cannot yet discover its purpose, its existence in some modification in all of these three very distinct forms, and in, as far as is known, no other mammal, is a strong corroboration of the view, formed upon other evidence, of their close affinity and common descent.

In the ass, the pouch is said to be deeper than in the horse, and areolated at the blind extremity; but detailed comparative observations upon its condition in the different species of existing *Equidæ* and upon its development in the horse are almost entirely wanting, and would well repay the trouble the investigation would cost. The nearest analogue in other orders of mammals is perhaps the singular pouch developed from the upper part of the commencement of the nasal passage of the "bladder-nosed" seal (*Cystophora cristata*), which the animal has the power of inflating with air when excited. The analogy is, however, by no means close, as in the seal the pouch is only found in the male, and not even in the young of that sex. If the sac in the horse is the remnant of some organ which formerly played a more important part in the economy of the race, we should expect to find it proportionately larger in the younger individuals of the existing species than in the adults.

GUTTURAL POUCHES.

Other equally mysterious structures are the "guttural pouches"— also diverticula of the respiratory passages—large cavities containing only air, one on each side, situated at the base of the skull behind the pharynx, and connected with the Eustachian tubes (the canals which convey air to the internal chamber of the ear) and which in the most approved works on veterinary anatomy are said to be "found only in solipeds." Exactly similar pouches exist in the tapir, but I am not aware whether they have as yet been looked for in the rhinoceros. They have been supposed to have some use in connection with the function of hearing; but it is possible they may rather be classed with the numerous large air sinuses found within the bones of the head of the horse, in common with most other mammals, the object of which is evidently to give increased volume without increased weight, and thus to furnish wide surface for the attachment of muscles and for the support and protection of various organs situated within the head.

It has been pointed out that in the artificial conditions under which some domestic horses live, these pouches may become sources of trouble and even danger. As they communicate with the nasal chambers by slit-like orifices, when the horse sniffs air is drawn into them. Should this occur when the animal is feeding in a manger or nose-bag, or the food is dusty, minute particles may enter the pouches and set up inflammation, or give rise to the formation of solid concretions. There is reason to believe that millers' horses are more liable to these affections than others.*

The Neck.

The skeleton of the horse's neck is formed of seven vertebræ, the same number as in man, and with very few exceptions, indeed, in all other mammals, whether the neck itself be short or long.

The skull is attached to the first vertebra of the neck, called the *atlas*, by a deep "ball-and-socket" joint, which allows of motion in various directions. The two projecting condyles of the skull together form the ball, which fits into the hollow front surface of the atlas. This vertebra also turns freely on the second, or the *axis*, and there is a certain but more limited amount of motion at each of the succeeding five joints. The combined action of these numerous joints permits of very free play to the head in all required directions.

* See J. Bland Sutton, *Evolution and Disease* (1890), p. 94.

The neck joins the front end of the *thorax* or chest, the skeleton of which is formed by the dorsal or thoracic vertebræ above, the sternum or breast-bone below, connected together by the hoop-like ribs. As seen in the figure (Frontispiece), the cervical or neck vertebræ are flat above, but those of the thoracic region have long processes projecting upwards, and forming together the ridge of the middle of the back. Those of the third, fourth, and fifth vertebræ, which are situated between the shoulder-blades, are the longest, and correspond externally to the "withers," the highest point of a horse's back, across which the measuring-rod is placed when taking his height.

The upper contour of the neck of the living horse is altogether different from that of the skeleton, the great depression seen in the latter in front of the high spines of the thoracic vertebræ being filled up in the middle line by a remarkable structure called the "cervical ligament," and on each side of this by large masses of muscles which raise and turn the head, and above all by the median "crest," a soft, but firm, fibrous, and fatty ridge immediately beneath the skin from which the mane grows.

The cervical ligament (*ligamentum cervicis, ligamentum nuchæ* or "pack wax") which in man is quite rudimentary, as his head, balanced on the top of the vertebral column, requires no special support, is immensely developed in the horse. It consists mainly of a strong elastic cord, which is attached in front to the upper part of the back of the skull (Fig. 22, O), and posteriorly to the elongated spines of the dorsal vertebræ (S). Between this, the *funicular* part of the cervical ligament, as it is called, and the bones of the neck, fibers of the same material, pass downwards and forwards, to be attached to the upper surface of all the different cervical vertebræ, except the atlas, forming a *lamellar* part, which lies in the middle line dividing the muscles of each side of the neck. This structure, though called a ligament for want of a better name, differs from the true ligaments, which connect bones together at the joints, in being of a yellow color and in having in a very marked degree the property of *elasticity*. It will bear considerable stretching, and then will return again to its normal length, which a true ligament composed of ordinary white fibrous tissue will not do.

Elastic material is often made use of in the animal economy to great advantage mechanically, restoring without effort to its proper position a part which has been temporarily disturbed from it, and thereby saving a vast expenditure of muscular power. The valves of an oyster or a cockle-shell are opened by an elastic hinge and closed by contraction of a muscle placed between them. Under the ordinary conditions of life it is necessary that they should remain open in order that the water containing air and nutriment may pass freely over the gills and mouth of

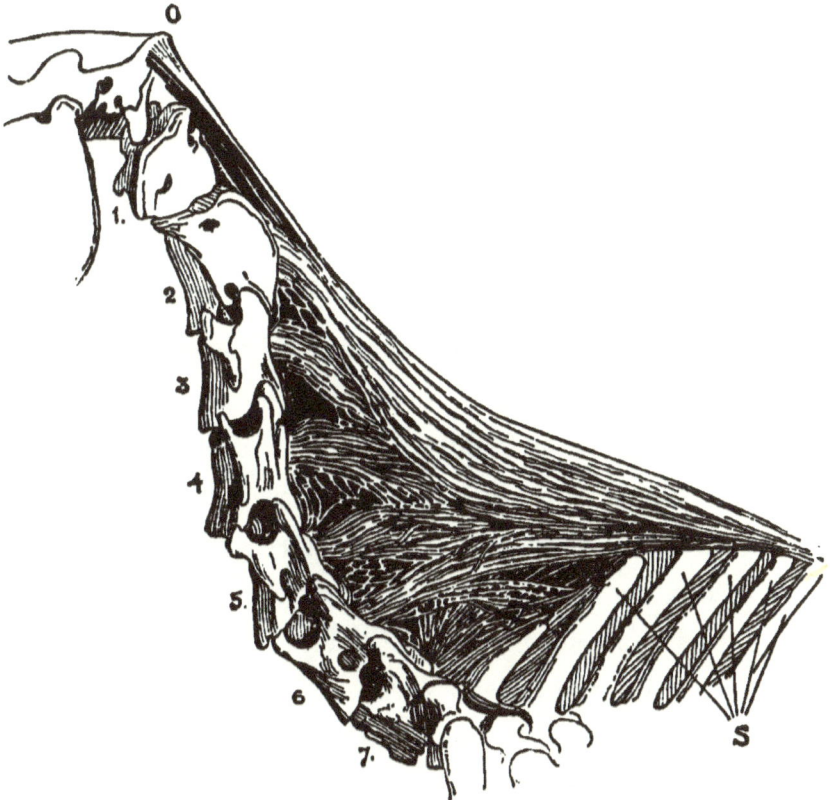

Fig. 22.—The cervical ligament, and bones to which it is attached (from Leisering). *O*, occipital crest of skull; 1 to 7, the seven cervical vertebræ; *S*, spinous processes of the anterior dorsal vertebræ.

the animal. They only need to be closed occasionally on the approach of some dangerous enemy. To close the valves and to maintain them in this position requires an effort; directly this effort is relaxed they open again by the simple mechanical effect of the elastic ligament. If they had to be opened and maintained in the open position, by a muscular effort, a far greater expenditure of power would be required on the part of the animal. In the same way in our own breathing, in which the air is alternately drawn in and expelled from the lungs by the expansion and contraction of the chest-walls, nearly half the labor, with corresponding expenditure of energy and waste of muscular tissue, is saved by the application of elasticity as the principal cause of the contraction which follows each muscular effort by which the act of expansion is performed.

As already mentioned, the horse's head, owing chiefly to the immense apparatus required to grind its necessary supply of food, is of great weight, and if it had to be supported at the end of the long neck entirely by a muscular effort on the part of the animal, great expenditure of force, requiring a still larger supply of food to keep it up, would take place. But, thanks to the structure, attachments and physical properties of the cervical ligament, it is no effort whatever to the horse to keep its head in the proper position. In fact, this ligament is so disposed, and of such strength and elasticity, as to allow the head to be moved up or down or from side to side, as required, by a very slight exertion of muscular action, but directly this ceases to return it to the position best suited for a state of repose.

Probably, if those who have to do with the harnessing of horses were better acquainted with this admirable mechanical apparatus for holding up the head in a natural and unstrained position, they would think it less necessary to supplement the cervical ligament by an external contrivance for effecting the same _object, called the "bearing-rein," which, however, not being elastic, never allows the head, even momentarily, to be altered in position; which is generally fixed so tightly as to interfere greatly with the natural graceful curve of the neck, one of the horse's chief beauties; and which, being attached at one end through the tender corners of the mouth, must, if short enough to effect the object for which it is used, be a continual source of pain or irritation to the animal.

Of the numerous petty cruelties practiced by man upon the domestic animals in obedience to the dictates of fashion or custom, or out of mere thoughtlessness, the use of the bearing-rein as a regular part of the harness of a carriage or cart-horse is one of the least excusable. We may, notwithstanding all the protests of the sensible, continue, from the same motives, to deform and injure our own feet by pointed shoes, and our own waists by tight lacing, but we

11

ought to extend more consideration to the comfort and welfare of the helpless animals, which, for our own advantage, we have taken under our care.*

* See *Bits and Bearing-reins, with observations on Horses and Harness*, by E. F. Flower, London, 7th edit. 1885.

CHAPTER IV.

THE LIMBS.

General characters of the limbs of vertebrated animals—Comparison of the skeleton of the fore limb of the horse with that of man—Comparison of the skeleton of the hind limb of the horse with that of man—The muscles of the limbs —The warts or callosities on the external surface of the limbs—The ergot or spur—The chestnuts, or mallenders and sallenders—The hoofs.

General Characters of the Limbs of Vertebrated Animals.

The body of all vertebrated animals consists of a main or *axial* portion, the "trunk," terminating anteriorly in the head, and posteriorly usually more or less prolonged into a tail. The skeleton of this part consists of the skull, the vertebral column, and the ribs and sternum or breast-bone. All the organs essential to life are contained in the axial part of the body, and in some animals, as serpents and a few fishes, it is the only part present. In the very large majority of animals, however, there are added appendages, called "limbs," mainly adapted for the purposes of locomotion, and which are attached to the trunk by the intervention of solid structures, commonly called in anatomical language "girdles." These are so called because the pair of them, when completely developed, nearly encircle the body; but it must be admitted that it is not a very happy expression, as, except through the intervention of the vertebral column, they never form complete circles, and very often the "semigirdles" of each side are widely separated both above and below.*

These girdles and the limbs which they support are never more than two in number on each side, and are almost always placed, the one near the front, and the other near the hinder end of the trunk.† The anterior girdle and limb are often called "pectoral," and the posterior "pelvic," from the regions of the body in which they are situated. Though in the large majority of vertebrated animals both pairs of limbs are present, either one or the other may be in a very rudimentary condition, or even alto-

---

* The semigirdles are sometimes called "arches," which is hardly more expressive, as, though the word means a segment of a circle in any position, the transverse position is now almost universally associated with it.

† In some fishes the ventral fin, which corresponds to the hind limb of most vertebrates, is placed below, or even anterior to the pectoral fin, or true front limb.

gether wanting, though in such cases some trace of a girdle is nearly always found.

The limbs belonging to the same region of the different sides of the body resemble each other in all essential particulars, being symmetrical paired organs. The anterior and posterior limbs have a general resemblance in plan, although always differing in certain details, these differences becoming more marked when the limbs have different functions to perform. Thus, as in birds, the fore limbs may be modified into wings for flight, while the hind limbs are only used for walking on the ground; in which case the fundamental resemblance of the two is very much masked. In the horse, as we shall see, where they are all used for the same purpose, standing, walking, or running on the ground, the fore and hind limbs are remarkably similar in construction, perhaps as much or more so than in any other animal.

The fore and hind limbs resemble each other mainly in being divided primarily into three segments: one proximal, or nearest the body and articulated with the girdle which carries it; one middle segment, and one distal or farthest from the body.

The proximal segment has in both cases a single bone forming its supporting axis, which bone is called the *humerus* or arm-bone, in the fore limb, and the *femur* or thigh-bone in the hind limb. The middle segment of both limbs has two bones lying side by side—the *radius* and *ulna* in the fore limb and the *tibia* and *fibula* in the hind limb. The joint by which the proximal segment is attached to the corresponding girdle is called "shoulder" in the fore limb, and "hip" in the hind limb; that between the first and second segments is called the "elbow" in the fore limb and the "knee" (the "stifle" in the horse) in the hind limb (see the skeleton of the horse and man in Frontispiece).

The distal or third segment is of more complex character. It constitutes in the fore and hind limbs respectively the "hand" and "foot" of man, the fore and hind "foot" of quadrupeds, or, in more precise anatomical language, of general application to all animals, the *manus* and the *pes*. Each of these consists of a group of small bones at its proximal end, forming the *carpus* or wrist in the fore limb and the *tarsus* or ankle in the hind limb. Beyond these it always has a tendency to divide up into a number of rays, called digits, fingers, or toes (see Fig. 1, p. 15).

Leaving out of consideration certain vestigial structures which are held by some anatomists to indicate the possibility of the former existence of a larger number of digits, no known mammal has more than five digits in each limb. For the convenience of description the digits are distinguished by the numerals I. to V., counted from the radial to the ulnar side in the fore limb, from the tibial to the fibular on the hind limb. They are also sometimes named —(I.) *pollex* or thumb (fore limb), *hallux* (hind limb); (II.) *index;* (III.) *medius;* (IV.) *annularis;* * and (V.) *minimus.* Though five is the complete number, one or more may be in a very rudimentary condition, or altogether suppressed. If one is absent, it is most commonly the first; next follows the fifth. The third is never lost, although either the second or fourth, or both, may be absent.

In both limbs the normal arrangement is that the carpus or the tarsus, as the case may be, supports five long bones placed side by side, called the *meta-podials* (or *metacarpals* in the fore limb, *metatarsals* in the hind limb), and to the end of each of these are three distinct bones called *phalanges*, except in the case of the *pollex* and *hallux*, which have only two. The terminal or distal phalanges of the digits are often specially modified to support the external horny covering usually present, called nail, claw, and hoof, according to its form and size, and hence are spoken of as the "ungual phalanges."

This portion of the limb, being usually more or less broadened and flattened, presents two surfaces

---

* Being in man the finger on which the ring is commonly worn.

and two edges and borders. The surfaces are *dorsal*, which in the ordinary position of the feet of most mammals is turned forwards or upwards (the " back " of the human hand), and *ventral*, or *palmar* in the fore limb or *plantar* in the hind limb, turned backwards or downwards. The edges are *external* (ulnar in the fore and fibular in the hind limb) and *internal* (radial in the fore and tibial in the hind limb).

The flexure between the middle and distal segments of the limb is called the " wrist-joint " and " ankle-joint " in the fore and hind limbs respectively in man, which correspond with those called the " knee " and the " hock " in the horse.

These are the essential characters in which the fore and hind limbs resemble each other. Of the differences many are merely adaptive to the different purposes to which they are put. The perfect efficiency of action, even in those that bear the closest resemblance, is secured by a partial rotation on its axis of each from the shoulder or hip, as the case may be, so that the outer side of the hind limb at the next joint comes to correspond with the inner side of the fore limb; but, owing to a second rotation in the middle segment in the latter, the last segments, or hand and foot, are brought again into corresponding positions in the ordinary walking attitude, the first (radial and tibial) digits being on the inside edge, and the fifth (ulnar and fibular) on the outside. Besides these differences, there are others, the signification of which is not so clear, constantly met with in the arrangement of the bones of the carpus and tarsus. Moreover, it may be noted that the joint between the first and second segments of the hind limb (knee-joint) has almost always a special bone (*patella* or knee-cap), which is wanting in the fore limb.*

This general description will include such different limbs as those of a man, a seal, a bat, and a horse, all formed on the same common plan, but all modified for the different purposes they have to fulfill. We must now treat in greater detail the peculiarities of the limbs of the horse, and to render them more intelligible another form is required for comparison. We will therefore take that with which we are all most familiar, and commence with a comparative account of the bones of the fore limb in man and in the horse.

COMPARISON OF THE SKELETON OF THE FORE LIMB OF THE HORSE WITH THAT OF MAN.

To begin with the shoulder-girdle. In the full-grown man this consists of two bones, the *scapula*

* For further description of the correspondences and differences of the bones of the fore and hind limbs, see the author's *Osteology of the Mammalia*, p. 361, 3d ed., 1885.

or "true shoulder-bone," or "blade-bone" (which is
itself composed in infancy, and in some animals permanently, of two separate bones, the scapula proper
and the coracoid), and the clavicle or "collar-bone,"
a strong curved bar, united at its outer end with the
scapula, and at its inner end with the sternum or
breast-bone.   The scapula is of complex shape, with
strong projecting processes.   In the horse (see Frontispiece) the humerus especially, so prominent a fea-
ture in the scapula of man, being scarcely visible.*
There is no trace of a clavicle.   The scapula and the
limb attached to it are not in any way joined to the
rest of the skeleton by bone, but only by the muscles
which pass from one to the other.   The trunk is, in
fact, only slung between the two shoulder-bones.

These differences are entirely related to the different use and motions of the fore limb in man and the
horse respectively.   In man the humerus moves at
the shoulder-joint in every direction.   It can be
swung round so that its outer end forms a complete
circle.   The muscles by which these actions are per-

* Wincza has recently shown that in the early embryonic
condition of the scapula of the horse this process is relatively
much larger than in the adult.   This is in conformity with
the general law that the young show the more generalized,
and the old the more specialized condition.   He was unable
to detect any sign of a clavicle. *Morpholog. Jahrb.* Bd. xvi.
(1890), p. 647.

formed require for their attachment outstanding ridges on the scapula. This bone, moreover, requires to have a certain degree of fixity, especially provision against its being driven too far inwards or outwards during the lateral action of the arms. This is provided for by its being connected to the sternum by the intervention of the clavicle. In the horse there is practically but one action at the shoulder, and that not a very extensive one—a fore-and-aft hinge scapula is a very much simpler bone, long, narrow, flat, with the processes much less developed, the acromion motion. The fore limbs are never crossed forwards across the chest, or thrown upwards behind the back, as with our arms, and hence there is no necessity for a clavicle, and the muscles which pass from the scapula to the humerus, though present, are developed in a very different degree.

Corresponding with the freedom and play of movement of the human arm and hand, the first bone of the limb proper, the humerus, in man is long and slender and has a large globular upper extremity or "head," which plays freely in the shallow, cup-like (glenoid), articular surface of the scapula, constituting a true ball-and-socket joint. In the horse, on the other hand, the humerus is comparatively short and stout,* and its movements are extremely limited. It

* The actual length of the humerus of an average-sized horse and man is almost identical, as seen in Frontispiece.

is, in fact, so short, and placed so nearly horizontally, and so covered up with muscles, that externally this segment makes no distinct appearance, being buried in the body or trunk, from which the limb only separates itself at the commencement of the second segment or elbow-joint, instead of at the shoulder, as in man.

The skeleton of the second segment or forearm in man consists of two bones placed side by side—(1) the ulna, which is connected with the humerus by a simple hinge-joint, allowing motion of bending (flexion) and straightening (extension) only in one plane, and (2) the radius, which turns or rotates in a peculiar way round the former, carrying the hand with it, and thus enabling the palm or the back of the hand to be turned uppermost at will—motions described as " supination " and " pronation." In the horse there is nothing of the kind; the radius is a strong bone of almost equal size at both ends, and the ulna is reduced to its upper part, which is firmly fixed to the radius, its only function being to strengthen the very perfect hinge of the elbow-joint behind. The hand is thus permanently fixed in the prone position, with its dorsal surface turned forwards. A flexible and revolving wrist-joint, though essential to the performance of the duties required from the human hand, would be quite incompatible with those needed from the corresponding part of the horse.

The consolidation of parts into a single supporting column, so conspicuous in the forearm, is carried
out to a still greater extent in the last segment of the
limb of the horse. The eight carpal bones of the human hand are, it is true (with one exception, the trapezium, the inner bone of the distal or lower row,
which supports the thumb), all present, even to the
pisiform, which projects backwards from the others
on the outer side of the wrist. These bones are,
however, more solidly compacted together than in
the human hand, the flat surfaces by which they come
in contact scarcely allowing a trace of movement between them. The metacarpus consists mainly of one
great bone, the "cannon-bone" of veterinarians, rep-
resenting the third or middle metacarpal of the human hand ($3m$, Fig. 6, p. 39). Lying on each side
of this, and generally in full-grown animals united
with it, are two smaller bones, the "splint-bones" of
veterinary anatomy ($2m$ and $4m$). These represent
respectively the second and fourth metacarpals of the
human hand. Above they have thickened heads,
which articulate in the usual manner with the carpal
bones; but below they taper off almost to nothing,
ending some way above the lower end of the great
middle bone. The part commonly called the "knee"
of the horse thus corresponds to the back of the wrist
of man, and everything beyond or below it corre-

sponds to the hand proper, the hinder surface being
the palm, long and narrow in the horse, as it is short
and broad in man.   As only one metacarpal bone is
fully developed, there is but one digit or finger, which,
as in man, has three bones (phalanges, p^1, p^2, and p^3,
Fig. 6), connected by hinge-joints, allowing only the
motions of bending or straightening backwards and
forwards.   The first phalanx is somewhat elongated,
the next very short, and the last (the ungual phalanx)
very broad and of a peculiar semilunar form.   These
bones are in veterinary anatomy called respectively
the "large pastern" or *os suffraginis*, the "small pastern" or *os coronæ*, and the "coffin-bone" or *os pedis*.
The joint between the metacarpal and the first phalanx is the "fetlock," that between the first and second phalanges the "pastern," and that between the
second and third phalanges the "coffin-joint."

There are several other small bones in the horse's
foot which must be mentioned, and which belong to
the group called "sesamoids," bones developed in tendons where they play over joints.   In the human
hand there is a pair of these over the palmar surface
of the metacarpo-phalangeal joint of the thumb, but
none are developed in the other digits.   In the horse
there are three, all also on the palmar surface (or behind in the natural position); a pair of nodular form
placed side by side over the metacarpo-phalangeal

articulation (Fig. 6, s), and a single large, transversely extended one (Fig. 6, s^1), called the "navicular" bone, behind the joint formed between the second and third phalanges.

In standing at rest in the natural position the forearm and the metacarpus are nearly upright, and the three bones of the digit or finger form a nearly straight line with them, but inclining forwards at the lower end. The third, or ungual phalanx, alone rests, through the intermedium of the hoof, upon the ground, and receives the whole of the weight of one quarter of the animal's body.

The main peculiarities of the skeleton of the fore limb of the horse are these: the absence of clavicle, the elongated, narrow and flat scapula, the short and obliquely placed humerus, the consolidated radius and ulna, the immensely developed middle metacarpal and its digit, and the suppression of all the others. Moreover, all the joints from the shoulder downwards are simply hinge-joints, allowing free fore-and-aft flexion and extension, but scarcely any movement in any other direction.

COMPARISON OF THE SKELETON OF THE HIND LIMB OF THE HORSE WITH THAT OF MAN.

The pelvic differs essentially from the pectoral girdle, inasmuch as it is firmly fixed to the trunk, to fit it for the more important part the hind limb takes in sustaining and propelling the body in walking and running.   Several of the vertebræ of this region are united into a solid block, the *sacrum*, to the sides of which the upper part of each arch or semigirdle is in the closest contact by a large flat surface, and firmly bound by strong ligaments.   The arches are, moreover, united to each other in the middle line below, without the intervention of anything corresponding to a sternum or clavicle.   On the outer side of each semigirdle is a deep round cup-shaped depression, the *acetabulum*, into which the head of the first bone of the limb proper is received, and which therefore corresponds with the glenoid fossa of the shoulder.   The joint at this position is the "hip-joint."   There is no essential anatomical difference in the construction of the "pelvis," as the whole girdle is called, in man and in the horse, each lateral half being in both originally composed of three distinct bones—the *ilium*, the *ischium*, and the *pubis*—which unite before the animal is full grown to form a solid mass, which has received from the old anatomists the curious name of *os innominatum*. The actual form of the bones presents considerable differences, the comparatively broad and basin-like pelvis of man relating chiefly to the adaptation of the body to the upright position.

The bone of the first segment of the limb proper is called the *femur* or thigh-bone. As in the corresponding bone of the fore limb, it is in the horse comparatively stout and short, and placed very obliquely, the lower end advancing by the side of the body, and being so little detached from it that the knee-joint appears to belong as much to the trunk as to the limb; a position altogether in contrast to that of the knee of man, separated from the body by the whole length of the elongated, free, vertically placed thigh (see Frontispiece). The bone itself has, in addition to the usual two rough processes near the upper end for the attachment of muscles (the *trochanters*) found in man and other mammals, a prominent compressed ridge, curving forwards, placed on the outer edge of the shaft of the bone, somewhat lower down than the other two. This, the so-called "third trochanter," as mentioned in the first chapter, is characteristic of all known Perissodactyles, and is also found in some rodents, but not in man or in mammals generally.

The second segment of the skeleton of the hind limb is represented in the horse almost entirely by the tibia. The fibula, indeed, is present, and a distinct bone, but only appears as a slender styliform rudiment of the upper portion attached to the outer side of the tibia.

12

The third segment of the hind limb, the foot or *pes*, has undergone precisely similar changes from the generalized or typical form to those already described in the fore limb. In fact, below the carpal and tarsal bones (the "knee" and "hock" of the horse respectively) the fore and hind limbs are almost exact repetitions of one another. The great development of the third metatarsal bone, the rudimentary condition of the second and fourth, the complete absence of the first and fifth; the presence of only one digit, consisting of three phalanges, having almost precisely the same form (except that they are rather narrower in the hind than the fore foot), are common to both extremities. In this structure of the foot, especially in the possession of but a single toe on each limb, the horse is absolutely unique among mammals. A very small Australian marsupial (*Chœropus castanotis*) has but one functional toe (in this case the fourth), on the tip of which it walks, on the hind foot, but three other toes are present, and complete in all their parts, though very minute; and in the fore foot two nearly equally developed toes reach the ground.

As the first segment of the horse's hind limb is so much shorter proportionately than that of man, the last is as much longer, and being habitually carried in a totally different position has a very differ-

ent appearance. The backwardly projecting prominence in the hock of the horse corresponds to the
heel of man, and the hinder surface of the horse's
limb, from the hock to the hoof, corresponds to the
plantar surface or "sole" of the foot of man. Man
is "plantigrade," the whole of the sole of the foot,
including the heel, being placed on the ground in
standing; the horse is "unguligrade," walking only
on the hoof, incasing the tip (or last phalanx) of the
toe. Dogs and cats assume an intermediate position
("digitigrade"), for, although the metatarsal bones
and the heel are raised, not only the tips, but the
greater part of the plantar surface of the toes rests
on the ground.

The sesamoid bones of the hind foot exactly resemble those of the fore foot.

The Muscles of the Limbs.

Muscles are the organs by which all the movements of one part of the body in relation to any
other part are effected. They lie around the bones
and beneath the skin, giving the external form to the
animal, and constituting what is commonly called its
flesh.

Muscular tissue is composed of a great number
of exceedingly minute parallel fibers of peculiar
structure, and it differs from all other tissue in pos-

sessing the property of contracting in length (with corresponding dilatation in width) on the application of a stimulus, usually conveyed to it through the nerve the terminal fibers of which are distributed through it.   The electric current, or mechanical irritation, will act as a stimulus to contraction, but in the living state the will of the animal, conveyed from the brain along the nerve to the muscle, is the usual cause of action.   If the nerve is divided anywhere in its course between the brain and the muscle, the latter will no longer act in obedience to the will, and is said to be paralyzed, although it does not really lose its power of contraction, as may be proved by the application of any other appropriate stimulus either directly to the muscle or to the lower part of the divided nerve.

In order that muscles by their contraction may produce movements, they must be fixed by their two extremities to two different bones, which are connected to each other by a movable joint.   When the contraction brings the ends of the muscle nearer together than they were before the bones must follow, and their position in relation to one another must be changed.   It usually happens that one attachment of a muscle is to a point more fixed than the other, and this is then spoken of as its "origin"; the attachment to the bone that is most movable being called the "insertion." This distinction is, however, not always a satisfactory one, as most muscles may act on occasions either way. In the limbs, where the muscles lie more or less parallel to the long bones, it is convenient to speak of them as arising at the end nearest the body, and being inserted at that farthest from it. As a general rule this accords with their action.

The muscles are sometimes attached directly to the bone, or rather to the fibrous sheath (*periosteum*) which closely invests it, but very often, for obvious mechanical reasons, they are connected with the bones by the intervention of "tendons," strong non-elastic fibrous cords, which are fixed to the muscle at one end and the bone at the other. It is in the limbs especially that tendons play a prominent part, as it is far more convenient that many of the strong muscles that move the fingers and toes should not be placed close to the parts on which they act, as if they were they would give a very clumsy form to the limb. They are, therefore, situated higher up, near the body, where increased thickness and weight of the limb are no disadvantage, and they produce their effect on the toes through the intervention of long tendons, which run close down the side of the bone.

As all the joints of the limbs of the horse are simply hinge-joints, acting only in one plane, the muscles are almost all either simple "flexors," bending the distal segments backwards on the segment above, or "extensors," returning them to the straight position. The structure of the joints prevents the segments being bent forwards much beyond a straight line with the segment above. The extensors are placed upon the anterior or dorsal, and the flexors on the posterior or ventral surface of the limb.

In the human arm and hand there are muscles having many other functions, such as turning the hand round, spreading the fingers and bringing them in contact again, which, of course, are not required in the horse. In the limbs of all mammals having the typical number of five digits completely developed, the muscles, as might be supposed, are as numerous and arranged on much the same general plan as in man. It is, however, very remarkable that in the horse's limbs many more muscles exist than would be thought necessary for the very simple actions they have to perform. But it appears that the reduction of bones to a rudimentary condition, as in the case of the ulna and the fibula, or their entire loss, as in the case of four of the toes, has taken place more thoroughly than, and in advance of, that of the muscles which were originally connected with these bones, many of which linger, as it were, behind, though with new relations and uses, sometimes in a most reduced and almost, if not quite, function-less condition, and sometimes even with completely changed structure.

From this point of view the muscles of the horse's limbs form a most interesting study. It has been truly said by Dr. G. E. Dobson,* that if no other evidence were obtainable of his five-toed ancestors, the condition of the muscles of the foot would suffi-ciently indicate them.

In the fore limb, where the ulna is represented only by the olecranon (projection of the elbow) and a greatly attenuated upper part of the shaft, and the digits reduced to one, most of the forearm muscles of the five-toed mammals are represented, the proper extensor of the fifth digit (*extensor minimi digiti*) even surviving, although both its insertion and special function have been completely altered. In the hind limb the two flexors of the toes (*flexor digitorum lon-gus* and *flexor hallucis longus*) are both present, with well-developed tendons united in the foot as in the great number of five-toed mammals.

It must not, however, be supposed from what has just been said that anything like all of the numerous muscles that are developed in the hand of man, with its versatile functions, can be traced in the horse.

* "On the Comparative Variability of Bones and Muscles," etc. *Journal of Anatomy and Physiology*, vol. xix. p. 16.

That would be obviously impossible with such a reduction of the bony elements. The difference (far less marked in the upper part of the arm) is especially pronounced in the last segment or *manus*, or hand proper, where the fifteen intrinsic muscles of the human hand are represented by only five * in the horse. Four of these—the two *interossei* and the two *lumbricales*—are in a very greatly reduced condition; and the fifth, the short flexor (represented in the human foot by the muscle called the "first plantar interosseous"), is a remarkable instance of a structure not becoming rudimentary and useless, but, while retaining its size, position, and connections, being diverted from its original purpose and completely changing not only its function, but its structure. It is termed in veterinary anatomy "the suspensory ligament of the fetlock," and appears as a very strong band or cord of non-elastic fibrous tissue, lying close to the back of the large metapodial bone, attached above to the posterior surface of the upper extremity of this bone, and at its lower end dividing into two portions, which, diverging from each other, embrace the metatarso-phalangeal or fetlock joint, and are inserted partly into the sesamoid bones and partly into the extensor tendon on the dorsal aspect of the

---

* This is the number according to the usual statements, but recent careful dissections have shown traces of others.

first phalanx (see Fig. 25, 10, p. 191). The obvious mechanical use of this ligament (as it has now become) is to prevent over-extension of the fetlock-joint. If it is ruptured or stretched the animal becomes what is termed in veterinary language "broken down"—*i.e.*, the fetlock-joint sinks down, and the hoof has a tendency to tilt forwards and upwards.

"The most interesting point, however, in connection with this structure is that it bears its history on its face. Almost invariably two thin streaks of striated muscular fiber are to be found on its superficial surface, leading down to its two inferior divisions. Again, on examining its deep surface, two very distinct strands of pink fleshy tissue are always observed extending throughout the entire length of the ligament. These consist in each case of short, oblique, striated fibers converging towards the middle line of the ligament. They represent those muscular fibers of the two heads of the *flexor brevis* which have not yet been converted into fibrous tissue. On making a thin microscopic transverse section the muscular fibers are seen to sink deeply into its substance, but it is altogether so small in amount in comparison with the bulk of the ligament that it can exercise no function whatever." *

* D. J. Cunningham, *Zoölogy of the Voyage of H. M. S. Challenger*, Part XVI. Report on the Marsupialia, p. 95.

THE WARTS OR CALLOSITIES ON THE EXTERNAL SURFACE OF THE LIMBS.

The external covering, integument, or skin of the horse is generally smooth, thick, and tough; much thicker on the back, flanks, and exposed portions of the limbs, and thinner on the under and more protected parts. Like the same structure in all other mammals, it is composed of two very distinct parts: (1) An inner, thicker layer, made up of interlacing filaments of tough, fibrous tissue to which blood-vessels and nerves are abundantly distributed, and which also contain muscular fibers, and, in its deeper portions, small collections of fatty tissue, and everywhere numerous minute glands of two kinds, sudoriferous and sebaceous, the former secreting a watery fluid (the perspiration or sweat), and the latter an oily substance which lubricates the skin and hair. This layer is called the *derm* or *corium*. (2) Lying upon this, and formed as an exudation or secretion of its outer surface, is a layer called the *epidermis*, not sensitive, and without blood-vessels, soft and moist in its deeper, and therefore newly-formed strata, and hard and dry at its exposed surface. It is not fibrous, but composed of cells which are at first nearly spherical or polygonal, but gradually become flatter and more scale-like as they approach the surface. Over the greater part of the skin it forms an exceedingly thin layer, which, nevertheless, serves as a protection to the softer and more sensitive derm below; but in certain parts it accumulates in solid masses of various forms, constituting the hairs, horns, nails, claws, hoofs, etc. Wherever these great accumulations take place, the superficial part of the derm is specially modified so as to afford a larger vascular surface available for their production, being covered sometimes with ridges or *lamellæ*, but more often with more or less elongated conical or cylindrical projections called *papillæ*. Each hair grows on such a papilla, which is sunk in the bottom of a follicle or deep pit in the derm or true skin. Under whatever form it appears, the epidermis is continually being removed at the surface, flaking or peeling off in minute fragments, or being worn and ground away by the contact of external substances, or, as in the case of hairs, cast off entire. The loss is, however, compensated by the continual renewal of the tissue from the surface of the derm below.

The greater part of the limbs of the horse is covered by an even coat of short hairs, but on the hinder part of the last segment these are much elongated, and especially at the prominence behind the joint between the metapodial bone and the first phalanx of the digit, where they form a considerable tuft or lock, which has given the name of "fetlock" (*i.e.*, foot or feet lock) to this part of the horse's limb. The amount and coarseness of this growth of hair varies much with the breed of the animal. The prominence itself is formed partly by the sesamoid bones, but also, in the middle line, by a mass of dense adipose tissue (the "fatty cushion of the fetlock"). On the center or most prominent part of this can be seen, on both fore and hind limbs, when the hair around it is clipped off, a roundish, bare patch (Fig. 23, C, *b*, p. 179; Fig. 25, 19, p. 191), covered with a rough, thickened epidermis, called in French veterinary works the *ergot*, as sometimes the epidermis accumulates on it to such an extent as to produce an appearance comparable to a spur.

The area of this bare patch is relatively larger in the ass than in the horse.

I am not aware that the significance of this peculiarly modified and hairless spot of skin, with its fatty cushion beneath, has ever been pointed out; nevertheless, although generally not noticed at all, or dismissed in a few words, in all works on horse anatomy, it is, when properly understood, one of the most interesting features of the external and visible structures of the animal's body.

If we look at the palm of our own hand (which, as shown before, corresponds with the hinder sur-

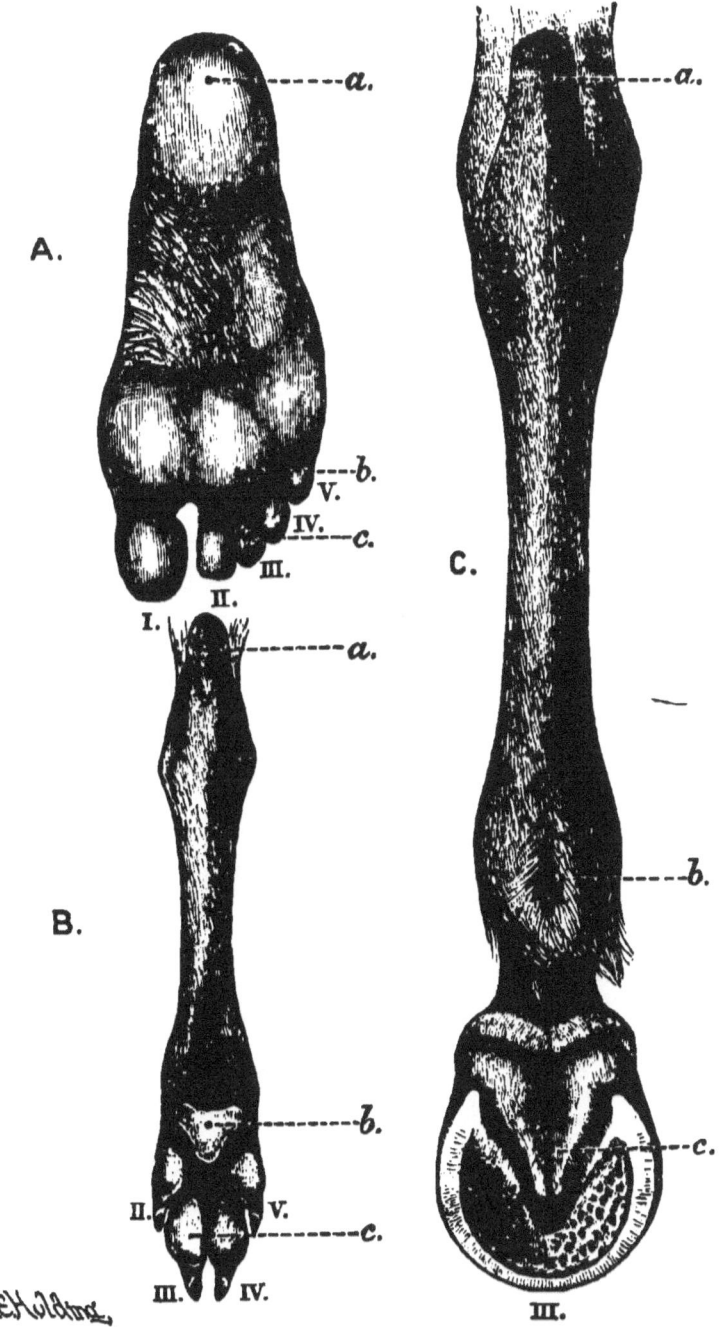

FIG. 23.—Plantar surface of the foot of—A, man; B,
dog; C, horse. The letters *a*, *b*, and *c* indicate
the corresponding points of the three. Compare
also Fig. 7, p. 47.

face of the fore limb of the horse below the so-called "knee"), we see slight prominences just behind the root of each finger and opposite the knuckles on the back of the hand, which mark the position of the joint between the metacarpal bones and the first phalanges of the digits. Over these, especially when the palm is subject to pressure and friction from hard manual labor, the epidermis is somewhat thickened. The sole of the foot presents exactly the same arrangement. In such an animal as a dog or a cat, in which this part of the foot comes to the ground in walking, there is a large trilobed prominent, bare pad (Fig. 23, B, *b*), composed of a thick fatty cushion, covered with a hardened epidermis, generally of a black color. There are also smaller pads in front of this on the under surface of each of the toes, but the large one corresponds with the coalesced three middle prominences of the human palm or sole just noticed.

In the horse's nearest living relatives, the tapir and rhinoceros, the same arrangement holds good. There is a large pad under the forepart of the middle of the foot, which in these animals rests on the ground, and also a hard sole under each toe (see Fig. 7, p. 47). Now the ergot of the horse clearly, both by structure and position, corresponds to the palmar or plantar pads of those animals which walk more or less on the palm and sole. Owing to the modified position of the horse's foot, standing only on the end of the last joint of the one toe, this part of the foot no longer comes to the ground, and yet the pad with its bare and thickened epidermic covering, greatly shrunken in dimensions and concealed among the long hair around, and now apparently useless in the economy of the animal, remains as an eloquent testimony to the unity of the horse's structure with that of other mammals, and its probable descent from a more generalized form, for the well-being of whose life this structure was necessary.

The ergot of the horse, placed in the middle line of the foot, must not be confounded, as has sometimes been done, with the parts bearing in French works the same name in the ox, and which are placed one on each side in a somewhat similar part of the foot. These are clearly shown by the structure of their horny covering, by the presence of bony elements within, and by comparison with their more developed condition in other ruminants, to be really the hoofs of the second and fifth digits, reduced to a very rudimentary condition.

Besides the ergot there are other patches, more obvious to ordinary observation, in which the skin is peculiarly modified from its usual structure. These are the so-called "chestnuts," or "mallenders" and "sallenders" as they are designated in old books.

They are patches on which no hair grows, but in which the papillæ of the derm or true skin are much enlarged and covered with an abundant and thickened epidermis, which becomes dry and horny and sometimes accumulates in considerable quantity on the surface, occasionally even making a horn-like projection. Their structure, in fact, is much like that of a wart or corn, but they are not the results of pathological changes, though often treated as such in old works on veterinary surgery. Even so enlightened a writer as Youatt includes them among diseases, and prescribes remedies both external and internal for the purpose of getting rid of them. They are, however, perfectly normal structures; they exist at birth, are equally developed in both sexes, and (allowing for certain limited individual variations) constant in form, size, and position. They constitute, moreover, one of the characteristic distinctions by which the species *Equus caballus* is distinguished from the other members of the genus.

They differ in form in the two limbs, but in both are placed upon the inner surface and nearer the hinder than the front border. That on the fore limb is *above* the carpal or wrist joint ("knee" of the horse), that on the hind limb *below* the ankle or "hock" joint. The former is about two inches long and three-quarters of an inch wide, pointed at each end, and lying obliquely, so that the long axis has its lower end directed backwards almost to the posterior border of the limb. When all the loose epidermis which incrusts it to a variable extent during life has been removed, the surface is seen to be elevated above the surrounding skin and to have definite prominent margins, and also to be generally convex, the tissue of which it is composed being thicker at the middle than at the edges. The hinder one is rather smaller and less elevated. Its posterior margin is nearly straight or regularly convex; its anterior margin is excavated in its upper third. It is therefore more pointed above than below. The upper end is about four inches below the point of the hock (*tuber calcis*). The natural color of both is dark slate, but when much dry epidermis collects on the surface they have a lighter or yellowish appearance.

In all the species of asses and zebras the hinder one is absent; but the one on the fore limb always exists, although in a modified form. It is broader or more oval in shape, and with a smoother and scarcely elevated surface. In the zebra it assumes the form of a large circular flat black patch of bare skin, nearly two inches in diameter.

The signification and utility of these structures are complete puzzles. Various suggestions have

13

been made, none of which will bear examination. One of the most plausible, especially in the light of modern comparative anatomy, is that they are rudiments or vestiges of the inner toe—the thumb or pollex of the fore limb, the great toe or hallux of the hind limb—which, as already shown, is not otherwise represented in the horse. There are, however, many objections to this theory. The inner toe is always the first to disappear in all mammals, and no traces of it are found in any ungulate, either Perissodactyle or Artiodactyle, except the most ancient forms. It is, therefore, most unlikely that anything of this digit should remain in the horse after the complete disappearance of the second, fourth, and fifth. In the next place, there is nothing beneath the modified patch of skin showing any trace of the structure of a toe, and the resemblance of this patch to a hoof is of the very slightest character, and, indeed, in the donkeys and zebras none whatever. But the most serious objection is the situation of the one that is most constant—that on the fore limb—where it is placed, not on the hand, as it would be if it represented the thumb, but upon the forearm, at some distance above the wrist-joint. Lastly, such a hypothesis is quite unnecessary, for they obviously belong to a numerous class of special modifications of particular parts of the cutaneous surface which occur in very many animals, the use of which is in most cases remarkably obscure. Bare spots, thickened patches or callosities, and tufts of elongated or modified hair, often associated with groups of peculiar glands, are very common on various parts of the body, but especially the limbs, of many ungulates, and to this category the "chestnuts" of the horse undoubtedly belong.*

If they teach us nothing else, they afford a valuable lesson as to our own ignorance, for if we cannot guess at the meaning or use of a structure so conspicuous to observation, and in an animal whose mode of life more than any other we have had the fullest opportunity of becoming intimately acquainted with, how can we be expected to account off-hand for the endless strange variations of form or structure which occur among animals whose lives are passed in situations entirely secluded from hu-

---

* The apparently capricious distribution of these may be illustrated by the following diagnoses of two groups or genera into which the pigmy chevrotains (small deer-like ruminants with some affinities to pigs) were divided by the late Dr. J. E. Gray, and which in all other respects closely resemble each other. (1) *Meminna.* "Chin entirely covered with hair. Hinder edge of the metatarsus covered with hair, with a large, smooth, naked prominence on the outer side rather below the hock." (2) *Tragulus.* "Throat and chin nakedish, subglandular, with a callous disk between the rami of the lower jaw, from which a band extends to the fore part of the chin. Hinder edge of the metatarsus naked and callous."

man observation, and of whose habits and methods of existence we know absolutely nothing?

THE HOOFS.

Any one who has read this book so far must be fully aware by this time that when we speak in ordinary language of a horse's "foot," the part we intend to designate is in reality the last joint of its toe.

As the value of the horse to man depends almost entirely upon its possessing this part in a sound and healthy state, it is one to which an immense amount of attention has been paid, and probably no other structure in the anatomy of any animal has been the subject of such minute investigation and elaborate description. It must be confessed that many of the current accounts of it are almost unintelligible, because the broad and interesting facts connected with it are completely obscured by a mass of minute, tedious, and unnecessary details, which seem to involve a comparatively simple organ in a cloud of mysterious technicalities. The fact is, that in all its main component parts, and in their relations to one another, the last joint of the toe of the horse precisely resembles that of any other animal, although some very remarkable and interesting modifications have taken place, adapting it for the special purpose it has to play in the economy of the horse.

The last segment or "joint" of the human finger (see Fig. 24, p. 190) differs as much from that of the horse in the use to which it is applied as is possible, yet an examination of its structure will afford a good key by which to understand the more complex arrangements of the latter. It contains one bone—the terminal or ungual phalanx. The proximal or upper end of this is wide transversely and hollowed out, fitting by a hinge-joint to the convex surface of the distal end of the second or middle phalanx of the digit. The two bones are firmly bound together by strong ligaments placed on each side of the joint, allowing free movement of flexion and extension, but not in any other direction. Below the joint the bone is somewhat constricted, but broadens out again into a sort of spoon-shaped end. The upper or dorsal surface is convex, the under or palmar surface flat. The ends of two tendons, which are worked by muscles situated a long way up in the limb, are fixed, one on the upper and the other on the under surface of the bone, and by their alternating contractions and relaxations cause it, with the structures around, to move in either direction on its hinge-like articulation. Between the bone and the skin are various soft structures, the terminations of arteries, veins, lymphatics, and nerves, embedded in a web of cellular or areolar fibrous tissue, with a considerable amount of fat, of which there is a special accumulation, forming a rounded pad, on the under surface of the end of the finger, called the bulb. In the skin over this part the sense of touch is especially developed.

The external surface is completely covered with a general continuation of the skin of the rest of the limb, the structure of which has been briefly described at p. 176. A part of this covering has, however, undergone a special modification to form the nail, which is a hard protecting shield for the most exposed part of the finger, and the freely-projecting edge of which serves many useful purposes. The nail is nothing more than a flattened plate of dry, hard, and horn-like epidermis, growing from a semilunar groove in the derm and from a depressed surface in front of this. This surface is covered, for the purpose of increasing its area, with slightly raised parallel longitudinal rows of papillæ, indications of which may be seen in the longitudinal ridges with which the surfaces of most nails are marked. The nail continually grows at the base and from its inner attached surface by the exudation of fresh epidermic material, which pushes forward the older-formed portion. This, if left to nature, eventually wears or breaks away at the free edge. The portion of the derm from which the nail grows is called the "matrix." The nail of the human hand is, generally speaking, flat, but its surface has a considerable curve from side to side, and also, though in a less degree, in its long diameter.

In those animals which belong to the typically unguiculated or clawed groups the bone of the last phalanx is long, but compressed from side to side, curved, and pointed, and the two sides of the nail bend completely round, so that their edges nearly meet at the middle line of the back or under surface of the finger or toe. In this way the nail becomes converted into a claw, which forms a sheath surrounding the bone. Usually, however, especially near the base, the edges do not quite meet, and there is a groove between them, filled up by softer epidermic material. As it is important for the due fulfillment of the function of claws that they should not be blunted by contact with the ground in walking, they are, in their most perfect condition, raised at their ends, the toes resting on bare cushions or pads placed beneath the joint between the middle and ungual phalanx.

When the horny covering of the last phalanx is modified into a hoof, on which the animal habitually rests and walks, the bone is generally broader and shorter, and the nail is curled round into the form of a short cylinder obliquely truncated, but with its edges not meeting behind, and with the interval filled by a mass of epidermis of great thickness but some-what softer structure than the nail proper, which is distinguished as the "sole" of the hoof. This corresponds in position to the rounded end of the human finger.

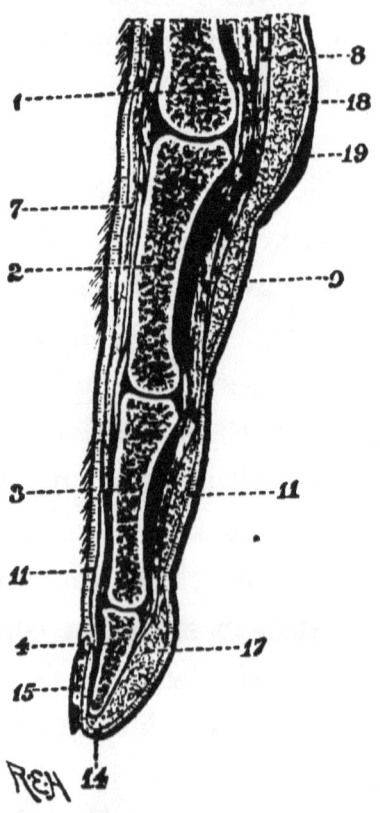

Fig. 24.—Section of the finger of man. 1, metacarpal bone; 2, first phalanx; 3, second phalanx; 4, third or ungual phalanx; 7, tendon of extensor muscle; 8, tendon of superficial flexor (*flexor perforatus*); 9, tendon of deep flexor (*flexor perforans*); 11 and 14, derm or true skin; 15, nail; 17, fibro-fatty cushion of end of finger; 18, ditto of palm behind metacarpo-phalangeal joint; 19, thickened epidermal covering of the same. The corresponding parts of Figs. 24 and 25 have the same numbers.

These three principal modifications of the same
structures, described respectively as nails, claws, and

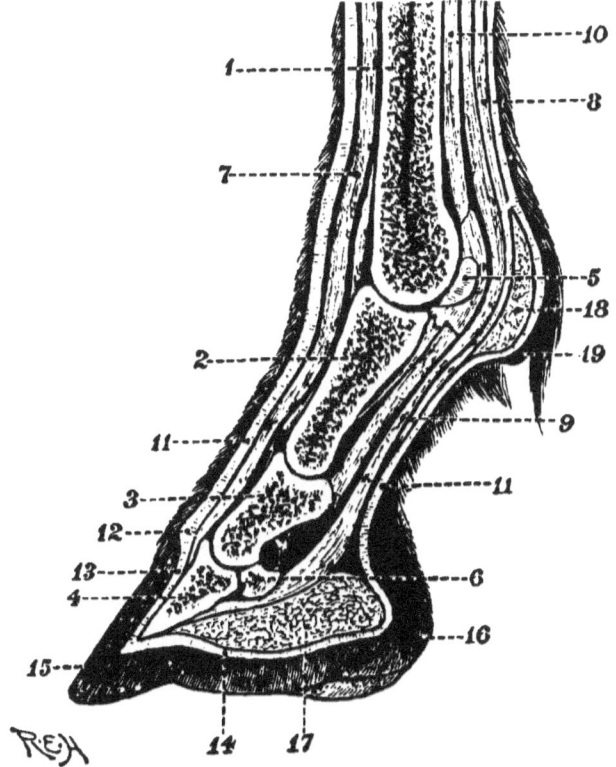

Fig. 25.—Section of the foot of horse. 1, metacarpal
  bone; 2, first phalanx; 3, second phalanx; 4, third
  or ungual phalanx; 5, one of the upper sesamoid
  bones; 6, lower sesamoid or "navicular" bone; 7,
  tendon of extensor muscle; 8, tendon of superficial
  flexor (*flexor perforatus*); 9, tendon of deep flexor
  (*flexor perforans*); 10, short flexor or suspensory
  ligament of the fetlock; 11, derm or true skin, continued into 12, coronary cushion; 13, laminal, and
  14, villous portions of the hoof matrix; 15, hoof;
  16, the "heel"; 17, plantar cushion; 18, fibro-
  fatty cushion of the fetlock; 19, bare patch with
  thickened epidermal covering or "spur." The corresponding parts of Figs. 24 and 25 have the same
  numbers.

hoofs, are met with throughout the Mammalian class, with numerous varieties or modifications of each, and transitional conditions by which they pass one into the other.

The horse shows the most extreme development of size and perfection of structure to which the hoof has attained, even considerably exceeding in this respect his nearest living allies, the asses and zebras. The bone which constitutes its support resembles that of man in the way it is jointed on to the bone above by a transversely extended concavity, and in having the ends of the long tendons of an extensor and flexor muscle inserted into it, one on the anterior and one on the posterior surface; but the bone is wonderfully different in shape, being very short, greatly expanded laterally, and ending below in a sharp but wide, nearly semicircular, distal edge. The bone is remarkable for its dense, almost ivory-like character, and is channeled and perforated to allow the passage of blood-vessels. The presence of a large sesamoid bone (the "navicular" of veterinarians) behind the articulation between the middle and distal phalanges is related to the great development of these bones, and to increasing the mechanical advantage of the flexor tendon which passes over it. Although not present in most mammals, it is not peculiar to the horse, being found, though on a

smaller scale, in other Perissodactyles. Not only is
the ungual phalanx larger, or at all events broader
in proportion to the rest of the digit, than in any
other mammal, but the parts around it are increased
to a still greater ratio, in order to give that firm basis
of support necessary when only a single toe reaches
the ground. In addition to its breadth, the toe is
prolonged backwards on each side into rounded
prominences with a deep indentation between them,
called the "heels" of the foot, as in comparing the
toe of the horse to the entire human foot they occupy
much the same position as the heel of the latter,
though, of course, they are in reality totally different
parts. In order to provide a support for this enlarge-
ment, the internal framework consists, in addition to
the bones, of certain accessory parts—viz., a pair of
fibro-cartilaginous masses, called the "lateral carti-
lages," one attached to each side or wing of the ungual phalanx and extending backwards towards the
heel, and a large elastic fibro-cellular and adipose
"plantar cushion" (Fig. 25, 17), occupying all the
median region below and behind the bone. The
former have nothing corresponding to them in man,
but the latter agrees in position and structure to the
fibro-fatty cushion of the bulb at the end of the human finger (Fig. 24, 17).

In the horse this "plantar cushion" is of great

size and importance. It is wedge-shaped; the narrow, pointed end, which is turned forwards and
reaches to the middle of the under surface of the
foot, causes the median triangular prominence called
the "frog." Posteriorly in the middle line is a deep
depression (the "median lacuna"), bounded on each
side by the "branches of the frog," which end in
rounded projections, the "glomes of the frog" forming the lower part of the heels. Blood-vessels,
nerves, lymphatics, and connective tissue make up
the rest of the structure of the toe, and the whole is
incased in a prolongation of the ordinary skin of the
limb, which, however, has undergone some very considerable modifications. At a sharply defined line
(the "coronet") which runs all round the foot, highest in front and becoming lower behind, where it
drops rather below the most prominent part of the
heels and dips into the lacuna, the hairy covering
altogether ceases, and a very thick epidermis takes
its place, completely incasing the whole terminal
part of the digit, as a thimble upon the end of a finger. In order to provide for the nutrition and continuous growth of this abundant epidermic covering,
the derm has acquired a greatly modified condition,
being very thick and vascular, and its surface is
everywhere immensely increased by folds, ridges,
papillæ, or villi. This vascular and sensitive struct-

ure, the "matrix of the hoof," "subcorneous integument," or "keratogenous membrane," as it has been
called, may be divided into three portions, differing
in position and structure:

1. A rounded, prominent ridge ("coronary cushion"), convex from above downwards (Fig. 25, 12),
constitutes the upper edge of the hoof-matrix, immediately contiguous to the hairy skin. It encircles
the front and side of the toe, descending on each side
behind and becoming continuous with the prominences called respectively the glomes, branches, and
body of the frog. Its surface is everywhere covered
with numerous and well-developed little thread-like
prolongations, which, if it is placed in water, float
out, and give the surface a velvety appearance. These
papillæ fit into corresponding tubular depressions in
the epidermic covering. From the coronary cushion
the base of the "wall" of the hoof (to be hereafter
described) grows; it therefore exactly corresponds
to that portion of the matrix of the human nail
which forms the bottom of the semilunar groove.

2. Below this, and separated from it by a narrow
whitish band, the membrane has a very different appearance. It closely covers the front and sides of
the bone as far as its lower margin, and posteriorly
on each side is continued round the hinder border of
each of the lateral wings, and turns forward to reach

almost the center of the under surface of the toe at the apex of the plantar cushion or frog. This part of the hoof-matrix (Fig. 25, 13), which from its structure is called the "podophyllous" or "laminal" tissue, is deeper from above downwards in front, and gradually gets shallower posteriorly, the incurved ends (which correspond to the "bars" of the horny hoof) thinning almost to nothing at their terminations. Instead of being covered with irregularly-scattered villi, like the coronary cushion and frog, its surface is raised into very numerous (five or six hundred altogether) longitudinal, parallel, fine leaf-like ridges or "lamellæ," all extremely vascular and sensitive, and being themselves covered on each side with numerous other finer ridges, set obliquely upon them—a most complex and delicate apparatus, enormously increasing the superficial area of the keratogenous membrane. This region of the matrix of the horse's hoof corresponds to the flat surface below the human nail, and the longitudinal ridges observed in the latter are obviously the same structures as the lamellæ of the horse's foot, only in a very slightly developed condition.

3. The third part of the matrix of the hoof is that portion of the vascular derm which covers the lower surface of the pedal bone, or the "sole" of the foot. It is crescentic in shape, and bordered all round by

the lower edge of the lamallar tissue. This, like the first, has a fine villous surface. The great size of this region is one of the peculiarities of the horse's foot. The covering of the sole is continuous posteriorly with that of the plantar cushion, which has also a villous surface.

Over the whole of this soft, sensitive, and highly vascular derm is, as before said, a very thick epidermic layer, which is distinguished by the name of "hoof." One of the properties of the horny material of which this is composed, which specially fits it for the function it has to perform, is that it is a non-conductor of heat. It is also moderately hard, tough, and elastic, and, like all epidermic structures, not sensitive itself, though it will transmit impressions through its tissue to the sensitive structures below.

From the foot of the dead animal the hoof may be removed entire by maceration or by immersion in hot water, when it will be seen to form a hollow box or case of somewhat complex form, its inner surface being exactly moulded on the parts around which it grows. Its general shape is that of an obliquely truncated cone, considerably higher in front than behind. As in the vascular surface beneath it, several distinct portions or regions can be distinguished. The densest and most important part is the "wall" or "crust," which exactly corresponds to the whole of the human nail, though differing in its much greater thickness and in the sides being not only greatly prolonged backwards, but also sharply bent inwards and forwards, forming the "bars" (Fig. 26, 5 and 6). The upper edge of the wall is hollowed to fit on to the coronary cushion. The inner surface is longitudinally furrowed by deep and complex grooves, which exactly correspond to the delicate lamellæ and laminellæ of the "podophyllous tissue." The new hoof continually grows from the coronary cushion above, and slides down over the lamellæ of the derm, receiving from them upon its inner side a certain amount of addition to its thickness in the form of soft epidermic cells, which afterwards harden and become incorporated in the general mass. The external surface of the wall when in a natural condition is smooth and shining, and appears to be made up of fine, closely arranged, parallel fibers, passing in a straight line from the upper to the lower margin. There are also not unfrequently transverse grooves or rings, indicating varying conditions of the matrix at the time of growth, especially marked in certain abnormal states of health.

The different regions of the wall have received technical names useful for descriptive purposes. The front part is called the "toe" (Fig. 26, 2, 3); the two sides, the outer and inner "quarters" (1, 2, and 4, 3);

the points where the wall suddenly bends inwards and forwards, the "buttresses" (1 and 4); the inner reflected ends, which nearly reach the center of the sole, the "bars" (5 and 6).

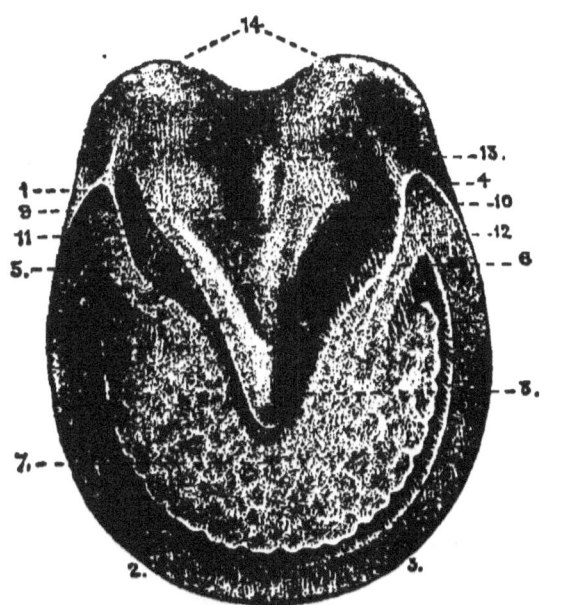

FIG. 26.—Under surface of hoof of horse (from Leisering). 1, 2, 3, and 4, the wall; the part between 2 and 3 being the toe; between 1 and 2, and 4 and 3, the outside and inside quarters; 1 and 4, the buttresses or angles of inflection of the wall to form the bars, 5 and 6; 7, the sole; 8, the point of the frog; 9 and 10, the branches of the frog; 11 and 12, the lateral lacunæ; 13, the median lacuna; 14, the heels.

The space between the lower edges of the wall is filled up in front by a flat or rather concave plate of a crescentic shape, called the "sole" (7), composed of softer and less fibrous material than the wall. Its

14

anterior and lateral borders, where it comes in contact with the inner surface of the toe and quarters of the wall, form nearly two-thirds of a circle. Its posterior concave border is bounded on each side by the bars, and in the middle it is deeply notched to receive the point of the frog.

Lastly, all the posterior part of the foot which comes to the ground is formed by the frog (8, 9, and 10) and its posterior prolongations, called the "branches" and "glomes," covered by a thick, callous, but not very horny epidermis, and which corresponds in form with that of the under surface of the "plantar" cushion previously described. We can distinguish a pyramidal median prominence (8), pointed in front where it reaches the center of the sole, with a groove on each side separating it from the bars, called the "lateral lacuna" (11 and 12); a deeper groove in the middle line farther back, the "median lacuna" (13), on each side of which are the branches of the frog, which posteriorly are swollen out into the glomes, rounded prominences forming the lower part of the heels (14), and continued round on each side of the hoof into the coronary cushion.

The terminal portions of the horse's four limbs are remarkably alike both in external appearance and internal structure, more so than are those of any other animal; and yet close inspection will show

differences by which they can be distinguished by a
practiced eye. The hoofs of the fore feet are broader
and rounder in front, those of the hind feet narrower
and more pointed. The right and left hoofs of either
limb can be distinguished by observing that the inner edge of the wall is flatter and the outer more
convex.

It will now be clearly seen that, in comparing the
under surface of the horse's foot with the tip of the
human finger, the free or lower edge of the wall of
the former corresponds with the free edge of the nail
of the latter, only vastly more developed in extent, in
complexity of involution, and in thickness; the frog
and all its accessory parts to the rounded free tip and
bulb of the finger, also greatly developed, so as to
form the heel-like projection so essential to give stability to the horse's foot in standing; while the sole
is only represented by the thin curved line between
the under surface of the nail and the skin covering
the tip of the finger.

Comparing the horse's toe with that of a clawed
animal—a dog or cat, for instance—the wall of the
hoof represents the horny sides of the claw; the sole
the narrow soft under surface of the claw, where the
edges do not meet; the frog and its branches and
glomes the smaller oval bare pad under the toe;
while the ergot or bare space behind the fetlock-joint

represents (as shown before) the large pad under the middle of the foot.

The hoofs of asses and zebras, though formed on exactly the same general plan as those of the horse, differ in being smaller, and especially narrower. The different parts of the inferior surface, the wall and bars, the sole and the frog, can be made out, though they are less distinctly marked from each other than in the horse.

The mechanical arrangement of the under sur-face of the horse's hoof in its natural state is admi-rably adapted to the purpose it has to fulfill. The different varieties of horny tissue of which it is com-posed and their complex arrangement recall those of the grinding-surface of the molar teeth. The wall or crust, completely encircling the front and sides, and reflected inwards and forwards almost to the center, being composed of a harder and more resist-ing material than the rest, like the enamel of the teeth, always stands out as a ridge beyond the other structures, and not only bears the principal weight, but prevents the tendency to slip which a uniformly smooth surface would have. The sole is more or less concave, being less dense and its surface exfoliating more readily than the other parts, but it comes in contact with the ground when this is of a soft and yielding nature. The projections formed by the elastic plantar cushion, covered by the horny frog and its backward prolongations, also bear much of the weight when these parts are left in their natural condition, though the ease with which they yield to the paring-knife offers a temptation which farriers seem rarely able to resist, much to the detriment of the proper action of the horse's foot.

In a state of nature, when the animal is free to choose the ground it runs over, the wear of the hoof is in exact proportion to its growth, and the organ always maintains itself in perfect condition. If, however, the horse is confined to ground so soft that only insufficient abrasion of the free surface of the hoofs can take place, they grow to an abnormal length. Horses turned out in the Falkland Islands, where the whole of the surface of the land consists of soft, moist, mossy bog, often have hoofs nearly a foot in length, bending and curling up in various directions, so that the animal at last can scarcely stand upon them. On the other hand, horses that are kept at work upon artificially hardened roads wear their hoofs so much faster than they grow, that from time immemorial their masters have found it necessary to protect them with some kind of artificial covering. Hence the almost universal use of iron shoes for horses in a state of domestication. Unfortunately, the subject of horse-shoeing has been too long left in

the hands of ignorant mechanics, by whose obstinate
adhesion to routine and ancient custom all the attempts of those who have endeavored to introduce a
more rational system are constantly foiled. This
subject, however, though of immense practical importance, is beyond the domain of natural history,
except in so far that a knowledge of the structure and
action of the foot in its natural state ought to be a
guide to those whose duty it is to counteract the un-
natural conditions to which we subject it.*

* Among many other works, see a small pamphlet on *The
Structure of the Horse's Foot, and the Principles of Shoeing*, by
Prof. G. T. Brown, C.B., reprinted from the *Journal of the
Royal Agricultural Society of England* (1888), and the larger
work of Dr. George Fleming, C.B. on *Horse Shoes and Horse
Shoeing*, 1889.

*A*PPLETONS' STUDENTS' LIBRARY. Con-

sisting of Thirty-four Volumes on subjects in Science, History, Literature, and Biography. In neat 18mo volumes, bound in half leather, in uniform style. Each set put up in a box. Sold in sets only. Price, per set, $20.00. *Containing:*

HOMER. By W. E. GLADSTONE. } 1 vol.
SHAKSPERE. By E DOWDEN. }

ENGLISH LITERATURE. By S. A. BROOKE. }
GREEK LITERATURE. By R. C. JEBB. } "
PHILOLOGY. By J. PEILE. }
ENGLISH COMPOSITION. By J. NICHOL. } "
GEOGRAPHY. By G. GROVE. }
CLASSICAL GEOGRAPHY. By H. F. TOZER. } "
INTRODUCTION TO SCIENCE PRIMERS. By T. H. HUXLEY. } "
PHYSIOLOGY. By M. FORSTER. }
CHEMISTRY. By H. E. ROSCOE. } "
PHYSICS. By BALFOUR STEWART. }
GEOLOGY. By A. GEIKIE. } "
BOTANY. By J. D. HOOKER. }
ASTRONOMY. By J. N. LOCKYER. }
PHYSICAL GEOGRAPHY. By A. GEIKIE. } "
POLITICAL ECONOMY. By W. S. JEVONS. } "
LOGIC. By W. S. JEVONS. }
HISTORY OF EUROPE. By E. A. FREEMAN. }
HISTORY OF FRANCE. By C. M. YONGE. } "
HISTORY OF ROME. By M. CREIGHTON. }
HISTORY OF GREECE. By C. A. FYFFE. } "
OLD GREEK LIFE. By J. P. MAHAFFY. }
ROMAN ANTIQUITIES. By A. S. WILKINS. } "
SOPHOCLES. By LEWIS CAMPBELL. } "
EURIPIDES. By J. P. MAHAFFY. }
VERGIL. By Prof. H. NETTLESHIP. } "
LIVY. By W. W. CAPES. }
DEMOSTHENES. By S. H. BUTCHER. } "
MILTON. By S. A. BROOKE. }

THE APOSTOLIC FATHERS AND THE APOLOGISTS. By Rev. G. A. JACKSON.
THE FATHERS OF THE THIRD CENTURY. By Rev. G. A. JACKSON.
THOMAS CARLYLE: His Life, his Books, his Theories. By A. H. GUERNSEY.
RALPH WALDO EMERSON, Philosopher and Poet. By A. H. GUERNSEY.
MACAULAY: His Life, his Writings. By C. H. JONES.
SHORT LIFE OF CHARLES DICKENS. By C. H. JONES.
SHORT LIFE OF GLADSTONE. By C. H. JONES.
RUSKIN ON PAINTING.
TOWN GEOLOGY. By CHARLES KINGSLEY.
THE CHILDHOOD OF RELIGIONS. By E. CLODD.
HISTORY OF THE EARLY CHURCH. By E. M. SEWELL.
THE ART OF SPEECH. Poetry and Prose. By L. T. TOWNSEND.
THE ART OF SPEECH. Eloquence and Logic. By L. T. TOWNSEND.
THE WORLD'S PARADISES. By S. G. W. BENJAMIN.
THE GREAT GERMAN COMPOSERS. By G. T. FERRIS.
THE GREAT ITALIAN AND FRENCH COMPOSERS. By G. T. FERRIS.
GREAT SINGERS. First Series. By G. T. FERRIS.
GREAT SINGERS. Second Series. By G. T. FERRIS.
GREAT VIOLINISTS AND PIANISTS. By G. T. FERRIS.

*A*PPLETONS' ATLAS OF THE UNITED STATES. Consisting of General Maps of the United States and Territories, and a County Map of each of the States, printed in Colors. Imperial 8vo. Cloth, $1.50.

The Atlas also contains Descriptive Text outlining the History, Geography, and Political and Educational Organization of the States, with latest Statistics of their Resources and Industries.

NEW YORK: D. APPLETON & CO., PUBLISHERS.

*T*HE ICE AGE IN NORTH AMERICA, and its *Bearings upon the Antiquity of Man.* By G. FREDERICK WRIGHT, D. D., LL. D. With 152 Maps and Illustrations. Third edition, containing Appendix on the "Probable Cause of Glaciation," by WARREN UPHAM, F. G. S. A., and Supplementary Notes. 8vo. 625 pages, and complete Index. Cloth, $5.00.

"Prof. Wright's work is great enough to be called monumental. There is not a page that is not instructive and suggestive. It is sure to make a reputation abroad as well as at home for its distinguished author, as one of the most active and intelligent of the living students of natural science and the special department of glacial action." —*Philadelphia Bulletin.*

*T*HE GREAT ICE AGE, and its Relation to the *Antiquity of Man.* By JAMES GEIKIE, F. R. S. E., of H. M. Geological Survey of Scotland. With Maps and Illustrations. 12mo. Cloth, $2.50.

A systematic account of the Glacial epoch in England and Scotland, with special reference to its changes of climate.

*T*HE CAUSE OF AN ICE AGE. By Sir ROBERT BALL, LL. D., F. R. S., Royal Astronomer of Ireland, author of "Starland." The first volume in the MODERN SCIENCE SERIES, edited by Sir JOHN LUBBOCK. 12mo. Cloth, $1.00.

"An exceedingly bright and interesting discussion of some of the marvelous physical revolutions of which our earth has been the scene. Of the various ages traced and located by scientists, none is more interesting or can be more so than the Ice age, and never have its phenomena been more clearly and graphically described, or its causes more definitely located, than in this thrillingly interesting volume."—*Boston Traveller.*

*T*OWN GEOLOGY. By the Rev. CHARLES KINGSLEY, F. L. S., F. G. S., Canon of Chester. 12mo. Cloth, $1.50.

"I have tried rather to teach the method of geology than its facts; to furnish the student with a key to all geology; rough indeed and rudimentary, but sure and sound enough, I trust, to help him to unlock most geological problems which may meet him in any quarter of the globe."—*From the Preface.*

*A*N AMERICAN GEOLOGICAL RAILWAY *GUIDE.* Giving the Geological Formation along the Railroads, with Altitude above Tide-water, Notes on Interesting Places on the Routes, and a Description of each of the Formations. By JAMES MACFARLANE, Ph. D., and more than Seventy-five Geologists. Second edition, 426 pp., 8vo. Cloth, $2.50.

"The idea is an original one. . . . Mr. Macfarlane has produced a very convenient and serviceable hand-book, available alike to the practical geo'ogist, to the student of that science, and to the intelligent traveler who would like to know the country through which he is passing."—*Boston Evening Transcript.*